State and Local
Climate and Energy Program

I0467999

LOCAL GOVERNMENT CLIMATE AND ENERGY STRATEGY SERIES

On-Site Renewable Energy Generation

A Guide to Developing and Implementing Greenhouse Gas Reduction Programs

Renewable Energy

U.S. ENVIRONMENTAL PROTECTION AGENCY
2014

EPA's Local Government Climate and Energy Strategy Series

The Local Government Climate and Energy Strategy Series provides a comprehensive, straightforward overview of greenhouse gas (GHG) emissions reduction strategies for local governments. Topics include energy efficiency, transportation, community planning and design, solid waste and materials management, and renewable energy. City, county, territorial, tribal, and regional government staff, and elected officials can use these guides to plan, implement, and evaluate their climate change mitigation and energy projects.

Each guide provides an overview of project benefits, policy mechanisms, investments, key stakeholders, and other implementation considerations. Examples and case studies highlighting achievable results from programs implemented in communities across the United States are incorporated throughout the guides.

While each guide stands on its own, the entire series contains many interrelated strategies that can be combined to create comprehensive, cost-effective programs that generate multiple benefits. For example, efforts to improve energy efficiency can be combined with transportation and community planning programs to reduce GHG emissions, decrease energy and transportation costs, improve air quality and public health, and enhance quality of life.

LOCAL GOVERNMENT CLIMATE AND ENERGY STRATEGY SERIES

All documents are available at: *www.epa.gov/statelocalclimate/resources/strategy-guides.html*.

ENERGY EFFICIENCY

- Energy Efficiency in Local Government Operations
- Energy Efficiency in K-12 Schools
- Energy Efficiency in Affordable Housing
- Energy-Efficient Product Procurement
- Combined Heat and Power
- Energy Efficiency in Water and Wastewater Facilities

TRANSPORTATION

- Transportation Control Measures

URBAN PLANNING AND DESIGN

- Smart Growth

SOLID WASTE AND MATERIALS MANAGEMENT

- Resource Conservation and Recovery

RENEWABLE ENERGY

- Green Power Procurement
- On-Site Renewable Energy Generation
- Landfill Gas Energy

Please note: All Web addresses in this document were working as of the time of publication, but links may break over time as sites are reorganized and content is moved.

CONTENTS

On-Site Renewable Energy Generation

EXECUTIVE SUMMARY

Developing and Implementing Renewable Energy Programs

A growing number of local governments are turning to renewable energy as a strategy to reduce GHGs, improve air quality and energy security, boost the local economy, and pave the way to a sustainable energy future. Renewable energy resources—such as solar, wind, biomass, hydropower, and landfill gas—reduce GHG emissions by replacing fossil fuels. Renewables also reduce emissions of conventional air pollutants, such as sulfur dioxide, that result from fossil fuel combustion. In addition, renewable energy can create jobs and open new markets for the local economy, and can be used as a hedge against price fluctuations of fossil fuels. Finally, local governments using renewable energy can demonstrate leadership, helping to spur additional renewable energy investments in their region.

Local governments can promote renewable energy by using it to help meet their own energy needs in municipal operations, and by encouraging its use by local residents and businesses. The renewable energy guides in this series present three strategies that local governments can use to gain the benefits of renewables: generating energy from renewable sources on-site, purchasing green power, and generating renewable energy from landfill gas.

On-Site Renewable Energy Generation

This guide describes a variety of approaches that local governments can use to advance climate and energy goals by meeting some or all of their electricity needs through on-site renewable energy generation. The sections in this guide discuss how local governments can work with utilities, local businesses, nonprofit groups, residents, state agencies, and green power marketers and brokers to plan and implement on-site renewable energy generation projects at local

RELATED GUIDES IN THIS SERIES

- **Renewable Energy:** *Green Power Procurement*

Green power is a subset of renewable energy that is produced with no GHG emissions, typically from solar, wind, geothermal, biogas, biomass, or low-impact small hydroelectric sources. Local governments can purchase green power for any remaining energy needs not covered by on-site renewable energy generation to help reduce their overall GHG emissions.

- **Renewable Energy:** *Landfill Gas Energy*

Landfill gas energy technologies capture methane from landfills to prevent it from being emitted to the atmosphere, reducing landfill methane emissions by 60–90%. Local governments can complement their landfill gas energy programs with other types of on-site renewable energy installations procurement to maximize the amount of their energy needs that are met by renewable sources.

- **Energy Efficiency:** *Energy Efficiency in Local Government Operations*

Local governments can implement energy-saving measures in existing local government facilities, new and green buildings, and day-to-day operations. Local governments can lead by example and take a holistic approach to reducing their GHG emissions by pursuing both energy efficiency and on-site renewable energy generation at their facilities.

- **Energy Efficiency:** *K-12 Schools*

Many local governments work closely with K-12 school district officials, who are often appointed by the local government executive or representative body. Because of this unique relationship, local governments are often well positioned to work through school districts to promote on-site renewable energy generation at schools in their communities.

government facilities and throughout the community. It is designed to be used by municipal energy coordinators, local energy and environment agency staff, environmental and energy advisors to elected officials, utility staff, and community groups.

Readers of the guide should come away with an understanding of the different types of on-site renewable energy technologies and applications, strategies for designing successful installations, and the associated financial considerations. The guide highlights examples of successful on-site renewable installations from across the United States to demonstrate how these technologies can help meet the diverse energy needs of communities of different sizes, governance structures, and locations.

The guide describes the benefits of on-site renewable energy generation (section 2); technologies and applications (section 3); key participants and their roles (section 4); the policy mechanisms that local governments have used to support on-site renewable energy generation projects (section 5); implementation strategies for effective installations (section 6); costs and funding opportunities (section 7); federal, state, and other programs that may be able to help local governments with information or financial and technical assistance (section 8); and two case studies of local governments that have comprehensive on-site renewable energy projects in place (section 9). Other examples of successful implementation are provided throughout the guide.

Relationships to Other Guides in the Series

Local governments can use other guides in this series to develop comprehensive climate and energy programs that incorporate complementary strategies. For example, local governments could combine on-site renewable energy generation with initiatives in **green power procurement**, **landfill gas to energy**, and **energy efficiency in K-12 schools** to help achieve additional environmental, economic, and social benefits.

See the box on page v for more information about these complementary strategies. Additional connections to related strategies are highlighted in the guide.

1. OVERVIEW

Many local governments are generating renewable energy at their own facilities and working with local businesses and residents to help them do the same at their offices and homes. By installing equipment that captures energy from sunlight, wind, water, and other renewable energy sources, local governments and communities can achieve substantial energy, environmental, and economic benefits. Installing on-site renewable energy generation systems at municipal facilities—and providing incentives to local businesses and residents to generate on-site renewable power—can also be an effective way to demonstrate a local government's commitment to meeting community GHG emission reduction goals.

RENEWABLE ENERGY AND GREEN POWER

Green power is a subset of renewable energy, and represents those renewable energy resources and technologies that provide the highest environmental benefit. Green power is produced from solar, wind, geothermal, biogas, eligible biomass, and low-impact hydro.

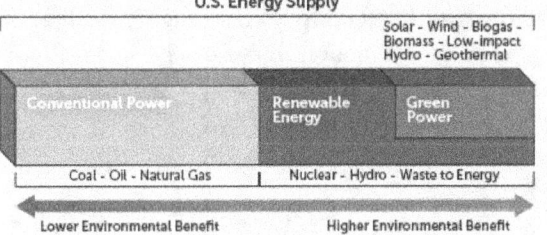

Green power sources produce electricity with an environmental profile superior to that of conventional power technologies, and produce no anthropogenic greenhouse gas emissions. EPA requires that green power sources must also have been built within the last 15 years in order to support "new" renewable energy development (U.S. EPA, 2007).*

January 1, 1997 is considered a definitive point in time when green power facilities could be adequately identified as having been developed to serve the green power marketplace. Green power facilities placed into service after January 1, 1997 are said to produce "new" renewable energy. The "new" criterion addresses the additionality requirement for the voluntary market (U.S. EPA, 2007a).

This guide highlights local government and community benefits associated with on-site renewable energy generation.[1] It also provides information on how local governments have planned and implemented on-site renewable energy generation activities at their facilities and throughout the community, and offers sources of funding and case studies. Links to more examples and resources are provided in Section 10, *Additional Examples and Information Resources* (see page 21).

2. BENEFITS OF ON-SITE RENEWABLE ENERGY GENERATION

On-site renewable energy generation can produce significant energy, environmental, and economic benefits by helping local governments and communities:

- **Reduce emissions of GHGs and other pollutants.** Substituting renewable energy for conventional energy can substantially reduce emissions of GHGs and other pollutants that result from local government activities. Fossil fuel combustion for electricity generation accounts for 67 percent of the nation's sulfur oxides (SOx) emissions, 23 percent of the nation's nitrogen oxides (NOx) emissions, and 40 percent of the nation's carbon dioxide (CO_2) emissions, pollutants that can lead to smog and acid rain, and increase the risk of climate change (U.S. EPA, 2008). Many local governments have developed plans with goals for reducing GHG emissions resulting from government and community activities. By generating renewable energy on site, local governments are demonstrating to their constituents that they are striving to meet these goals (U.S. EPA, 2004).

- **Hedge against financial risks.** On-site renewable energy generation systems can reduce local government energy costs by decreasing exposure to fossil fuel price volatility, which can lead to higher prices for grid-based electricity. This allows local governments to better anticipate and plan for future energy expenditures (U.S. EPA, 2004; AWEA, 2007).

---

[1] Generating renewable energy can provide a source of green power. Complementary information on how local governments can develop green power programs is available in EPA's Green Power Procurement guide in the Local Government Climate and Energy Strategy Series.

GENERATION CAPACITY AND PRODUCTION

Electricity production and consumption (measured in kWh) are functions of generation capacity (measured in kW) and time (measured in hours). System generation capacity depends on a site-specific capacity factor, which describes the system's actual energy output divided by the output that would be achieved if the system operated at full capacity. Electricity production can be calculated as follows:

Electricity production (kWh) =
Capacity (kW) x Capacity factor x Time (hours)

Solar photovoltaic panels typically have capacity factors between 0.21 and 0.26. For most utility-scale onshore wind turbines, the capacity factor is between 0.35 and 0.44.

As an example, the annual electricity production of a 10 kW wind system with a capacity factor of 0.15 would be calculated as follows:

10 kW x 0.15 x 8,760 hours = 13,140 kWh per year
(36 kWh per day)

Source: U.S. DOE, 2010

REDUCING GRID-BASED ELECTRICITY PURCHASES

Obtaining electricity from on-site sources can produce significant cost savings. A wind turbine with a generation capacity of 10 kW located at a site with average wind speeds of 12 miles per hour can produce approximately 10,000 kWh annually, enough to power a small building. Assuming an average price for commercial electricity of 10.7¢ per kWh (as of June 2013), the wind turbine would reduce annual grid-based electricity costs by approximately $1,070. With installed costs for on-shore turbines ranging between $1,200 and $2,100 per kW capacity and including a federal tax credit of 30%, these savings could mean a simple payback period of less than six years.

Sources: EIA, 2013; AWEA, 2007; NREL 2012b

In 2003, the city of Auburn, New York, installed a geothermal system to heat and cool its historic city hall at an installed cost of approximately $1 million, comparable to the cost of a conventional heating and cooling system. The geothermal system, which was installed in a way that blended with the historic building's internal and external architecture, was expected to save approximately $19,000 annually in operating and maintenance costs (including energy costs) over its lifetime due to expected increases in conventional energy prices (McQuay International, 2003). After the city hall project proved successful, Auburn converted other municipal buildings to geothermal, including the city's police and fire department building and the Cayuga-Onondaga Board of Cooperative Educational Services campus (City of Auburn and Cayuga County, 2009).

- **Support economic growth through job creation and market development.** Investing in on-site renewable energy generation can help stimulate local, state, and regional economies. On-site renewable energy generation systems require a considerable amount of raw materials, and purchasing these materials from local businesses can increase local manufacturing employment. Demand for construction, installation, and

maintenance of on-site renewable energy generation systems can create jobs and help develop the market for these technologies (NREL, 2012). Figure 1 on page 3 illustrates the direct and indirect jobs, earnings, and economic output derived from PV and large-scale wind projects funded by the U.S. Department of the Treasury's 1603 Grant Program.[2]

- **Demonstrate leadership.** Generating renewable energy at local government facilities can be an effective and visible way of demonstrating environmental leadership to the public.

In 2003, Lenox, Iowa, purchased and installed a 750 kW wind turbine to produce electricity for its own facilities at about the same time that the town's municipal electric utility began offering customers the option to purchase renewable energy. When nearly 13 percent of the city's households enrolled to purchase renewable-generated electricity, significantly higher than the national average of 1 to 3 percent, the city attributed the success of the program to the increased public awareness generated by the new turbine (Energy Services Bulletin, 2004).

Installing renewable energy generation systems at facilities that are visited frequently by the public can lead to greater community awareness of local government leadership and the benefits of clean energy activities.

---

2 *For more information on the 1603 Treasury program:* http://www.treasury.gov/initiatives/recovery/Pages/1603.aspx

MILWAUKEE: GAINING STEAM THROUGH SOLAR
INSTALLATIONS

Milwaukee, Wisconsin, has demonstrated leadership in
implementing solar power at the local level. The U.S.
Department of Energy's listed Milwaukee as one of 25
Solar America Cities in 2008, and the city has continued
to expand its solar programs since then.

In 2011, the city launched the Milwaukee Shines program
to help finance its use of solar energy. As part of this
program, the city incentivized Milwaukee-based solar
manufacturing facilities to sell affordable solar products
to local certified installers, reducing costs for customers.
The city also provided low-interest loans for residents
and businesses to finance solar installations.

Under the Milwaukee Shines program, the city has
already completed 50 solar projects and has set an
ambitious goal to install 1 MW of solar energy. As of late
2013 it had already installed 350 kW of that planned
capacity.

Source: City of Milwaukee, 2013; U.S. DOE, 2013

Highland Beach, Maryland, for example, is
demonstrating leadership by generating 100
percent of the energy used by its town hall
from renewable resources. The town hall uses
geothermal energy to reduce heating and cooling
loads, and solar photovoltaic panels produce

enough power to meet the balance of the building's
energy needs (Highland Beach, 2006).

- **Improve power quality and supply reliability.** Electricity has high "power quality" when the required
amount of energy is delivered consistently without
variation. Sags or spikes in voltage may be evidence of
poor power quality. The complex network of interconnections involved in generating, transmitting, and
delivering grid-based electricity causes temporal varia-
tions in the characteristics of delivered power. Because
on-site renewable energy generation systems have fewer
interconnections (e.g., transmission substations), electricity from these sources is likely to have higher power
quality than electricity delivered through the grid from
fossil fuel sources (U.S. EPA, 2004; MTC, 2002).

In addition to power quality, local governments want
to ensure the reliability of their energy supply. Disrup-
tions in energy supply can be a serious risk for local
governments, many of which own hospitals, schools,
and other facilities that house residents who may
rely on a consistent electricity supply. By installing
renewable energy generation systems on site, local
governments can improve energy supply reliability
and protect against grid-based electricity shortages or
blackouts (U.S. EPA, 2004).

FIGURE 1. JOBS, EARNINGS, AND ECONOMIC OUTPUT FROM PHOTOVOLTAIC AND LARGE
WIND PROJECTS

Summary Estimates of the Direct and Indirect Jobs, Earnings, and Output Supported			
	Average Jobs per year (FTE/year)	Total Earnings (Billions $)	Total Economic Output (Billions $)
During Construction Period (2009-2011)			
Large Wind	44,000–66,000	$7.7–$12.0	$23.0–$39.0
Photovoltaic	8,300–9,700	$1.5–$1.8	$3.5–$4.7
Total Direct + Indirect	52,000–75,000	$9.2–$14.0	$26.0–$44.0
During Operational Period (annual for system lifetime)			
Large Wind	4,500–4,900	$0.26–$0.29	$1.60–$1.70
Photovoltaic	610–630	$0.04	$0.09
Total Direct + Indirect	5,100–5,500	$0.3–$0.3	$1.7–$1.8
Source: NREL, 2012a .			

3. ON-SITE RENEWABLE ENERGY TECHNOLOGIES AND APPLICATIONS

Local governments can select from a range of technologies for on-site renewable energy generation. Renewable energy sources that can be captured using on-site systems include:

- **Wind.** Wind energy, which is captured on site using wind turbines, can be very cost-effective in areas with adequate wind resources. A 3 kW turbine with a 60- to 80-foot tower could reduce a facility's monthly electricity bill by 30 to 60 percent, assuming monthly electricity costs range between $73 and $115 (approximately 700 kWh to 1,100 kWh).[3] Using the national weighted average installed cost for sample wind projects in 2012 (approximately $1,940 per kW capacity), these cost savings could result in a simple payback period as short as six years (U.S. DOE, 2013a).

HULL, MASSACHUSETTS – WIND POWER

In December 2001, Hull, Massachusetts, a coastal town on a peninsula south of Boston, purchased a 660 kW wind turbine to replace a pre-existing structure that had once served the town's high school. Within its first two years, the turbine produced nearly 3,000 MWh of energy, demonstrating a capacity factor of 27 percent. A second turbine, commissioned in May 2006, has a capacity of 1.8 MW. Combined, the two turbines generate enough electricity to supply 11 percent of Hull's load.

The electricity from the turbines is generated at a cost of 3.4¢ per kWh, which is less than half of the 8.0¢ it would cost from the grid.

The town of Hull is now looking into offshore wind projects.

Source: Hull, 2008; Hull Municipal Light Plant, 2013.

As opposed to large utility-scale wind farm turbines, which can reach capacities as high as 3 MW, "small wind" turbines (turbines that have capacities of 100 kW or less) are often better suited for local facilities (AWEA, 2007a).[4] Wind turbines are most often installed in non-urban areas because installations typically require at least one acre of land and wind speeds averaging 15 mph at 50 meters above the ground (U.S. EPA, 2004). However, small turbines can be appropriately sited in urban areas.

In 2012, the oceanfront city of North Myrtle Beach, South Carolina, finished installing its second small-scale wind turbine. Together, the two turbines generate about 4 kilowatts of electricity to power a concession stand and a water slide on the beach. Rather than generating significant amounts of electricity, the project is meant to provide data to inform future off-shore wind development opportunities (Carolina Live, 2012).

- **Solar.** Sunlight provides an abundant source of renewable energy. Solar technologies take advantage of the sun's energy using two different capture methods: active and passive. Active solar technologies use complex mechanized collectors, such as photovoltaic panels, to collect and store solar energy. Passive solar technologies are less complicated and rely on the design and orientation of the collector rather than mechanical devices to absorb and store the sun's energy (EPA 2012). Technologies that use these methods include:

  - *Photovoltaics (PV).* PV systems directly convert sunlight into electricity using solar cells. These systems, which can produce electricity even in the absence of strong sunlight, can generate significant quantities of electricity depending on several factors, including quality of the sunlight and the system's mounted pitch. For instance, the San Diego Regional Energy Office estimates that PV systems in the San Diego area can produce between 1,400 kWh and 1,700 kWh per kW capacity annually (SDREO, 2007). The New York State Energy Research and Development Authority (NYSERDA) estimates that PV systems can produce between 1,000 kWh and 1,300 kWh per kW capacity annually in New York (NYSERDA, 2007). A 10-kW system that produces 1,500 kWh per kW capacity per year could thus produce 15,000 kWh annually.

    PV systems are often installed on rooftops, making them ideal for local government buildings in areas where open space is limited. Local governments have installed PV systems at fire stations, libraries, and a wide range of other buildings. PV systems can also be installed as stand-alone systems (i.e., systems that are not connected to the electricity grid) on parking meters, bus stop canopies, and on parking lot lights (Portland, 2007; Phoenix, 2007; Anaheim, 2001).

---

[3] KWh approximations determined using most recent average retail price for conventional electricity (10.47¢ per kWh) (EIA, 2013a).

[4] Most small wind turbines have capacities of less than 25 kW (AWEA, 2007a).

TALLAHASSEE, FLORIDA – SOLAR INITIATIVES

The city of Tallahassee, Florida has installed PV and solar water heating equipment at multiple city facilities. A 10 kW PV system has been installed at a public gymnasium and aquatic center to provide up to 14,000 kWh annually. In addition, the city has installed an 18 kW PV system at the Capital Center Office Complex and a solar hot water system at City Hall. The solar hot water system captures heat from sunlight, concentrating it to heat water that is distributed throughout the facility.

In addition to installing renewable energy generation equipment at local government facilities, the city offers rebates of up to $450 for solar water heating system installations at residential or commercial facilities. The city also offers low-interest loans to further ameliorate the costs of installing the equipment.

Source: Tallahassee, 2007, DSIRE, 2012.

> *Solar hot water.* Solar hot water technology uses sunlight to heat water in a collector and then distribute the heated water throughout a building, reducing a building's reliance on a conventional hot water heater that uses non-renewable sources of energy. Solar hot water devices can use either passive or active systems, although most are active (NREL, 2007a).

> *Solar process heating and cooling.* Solar process heating uses sunlight to provide space heating in buildings. This technology captures heat from sunlight using contained air or fluid as the medium. The captured heat is then fanned or pumped to distribute it throughout the building. The heat from a solar collector can also be used to cool a building in the same way that electricity is used to power air conditioning units (NREL, 2007).

▪ **Geothermal.** Geothermal systems capture the earth's heat for use in generating electricity and providing heating and hot water. In direct use applications, steam from beneath the earth's surface can be used to power turbines to produce electricity. This type of geothermal application is dependent on the availability of adequate geothermal reservoirs (reservoirs of water with temperatures between 68o F and 302o F), which are more common in the western United States.

A second type of geothermal technology uses heat pumps to capture the earth's natural heat to warm liquid that is pumped into buildings from underground piping to provide central heating or to heat water. In warmer seasons, geothermal heat pumps can exchange warm surface air for cooler below-ground air (U.S. DOE, 2006). Geothermal heat pump systems are typically installed at shallow depths (e.g., 4 to 6 feet below the surface). Because shallow ground temperatures are fairly constant throughout the United States, geothermal heat pumps can be effective in most locations (U.S. DOE, 2007a).

▪ **Biomass.** Electricity-producing steam turbines can be fueled by burning solid biomass feedstocks, such as plant material, construction wood, agricultural wastes, sewage, and manure. Biomass can be used to generate electricity by heating feedstocks in an oxygen-free environment to convert them into combustible oil or gas biofuels. This gasification process can be up to two times more efficient than burning solid biomass, and results in reduced GHG emissions. By siting biomass operations in areas that have abundant biomass resources, such as agricultural or forestry waste, local governments can take advantage of material that would otherwise be wasted (U.S. EPA, 2000; 2004).

The city of Battle Creek, Michigan entered into a contract to install an $18 million biomass gasification system for the U.S. Department of Veterans Affairs Medical Center (VAMC). The system, scheduled to go into operation in late 2013, will burn woody biomass products (such as wood chips, pallets, and other locally available residual wood wastes) to generate 2MW of renewable electricity and 14,000 pounds per hour of steam to heat the medical center. This project will also help the Department of Veterans Affairs reach its 2020 GHG reduction goals as well as comply with President Obama's Executive Order 13514, which requires that the federal government reduce its GHG emissions by 28 percent by 2020. This project will allow the VAMC to reduce its annual GHG emissions by 14,000 metric tons, reducing its carbon footprint by roughly 80 percent (PR Newswire, 2011).

- **Landfill gas and other biogas.** Equipping landfills and other facilities (e.g., wastewater and manure treatment facilities) to capture biogas provides a source of renewable energy from a byproduct that would otherwise be wasted. Biogas contains methane, a natural byproduct of anaerobic digestion of landfill refuse, sewage, and other products, which can be converted into electricity through conventional combustion processes. For example, a single landfill gas recovery project can reach capacities as high as 4 MW (U.S. EPA, 2004).

CITY CAPTURES METHANE AT MUNICIPAL WATER RECLAMATION FACILITY

The Truckee Meadows Water Reclamation Facility is jointly owned by the cities of Reno and Sparks, Nevada, and is operated by the city of Sparks. In addition to treating wastewater for reuse, the facility captures methane produced in the treatment process. The methane is used to fuel a 700 kW generator, which produces electricity that is sold to the local utility. In February 2007, Sparks received $287,000 from the utility in compensation for three years worth of electricity contributions to the grid.

Source: Green Jobs, 2007.

In 2013, the Cayuga County Soil and Water Conservation District opened a community digester facility that uses untreated manure from local farms to fuel an anaerobic digester. The digester produces biogas, which is combusted to achieve up to 633 kW of electricity generation capacity for county buildings. The byproduct from the digesters is returned to the local farmers in the form of a liquid fertilizer that contains less phosphorus and has a smaller pollution potential than raw, untreated manure (Cayuga County Soil and Water Conservation District, 2013).

Methane is a potent GHG that has a global warming potential 21 times that of CO2, and landfills are responsible for 17 percent of the nation's methane emissions (U.S. EPA, 2011). Landfill gas and other biogas recovery projects can contribute significantly to reducing the risks of climate change. For example, manure capture and utilization for biogas can reduce methane emissions from manure biodegradation by 2.75 metric tons of CO2 equivalent per cow per year

(U.S. EPA, 2004a). Using manure biogas to produce electricity can offset 0.9 metric ton of CO2 emissions per cow per year, by replacing grid-based electricity generated from conventional fuel sources (U.S. EPA, 2004a). These projects can produce other environmental benefits, including reduced waste odors and pathogens, as well as economic benefits (U.S. EPA, 2006b). A 3 MW landfill gas project, for example, can support more than 70 full-time jobs over the course of a year (U.S. EPA, Undated). For more information on landfill gas projects, see EPA's *Landfill Gas Energy* guide in the *Local Government Climate and Energy Strategy Series*.

- **Low-impact hydropower.** Hydropower projects capture the kinetic energy of moving water to produce electricity. While hydropower is renewable and produces relatively few GHG emissions, hydropower projects can have other impacts on the environment, such as obstructing fish passage and altering land resources by impounding excessive nutrients (U.S. EPA, 2006a). The Low-Impact Hydropower Institute (LIHI) provides certification to hydropower projects that demonstrate minimal impact on the environment. The EPA Green Power Partnership only recognizes as green power hydroelectricity that is generated by LIHI-certified projects (LIHI, 2008), run-of-river facilities equal or less than five MW of nameplate capacity, or facilities that consist of a turbine placed in a pipeline or irrigation canal.

SEATTLE, WASHINGTON – LOW-IMPACT HYDRO FACILITY

Seattle City Light, the municipal electric utility in Seattle, Washington, owns and operates the Skagit Project on the Skagit River. This LIHI-certified hydroelectric facility has a capacity of 690 MW and generates 2.5 million MWh annually. When conferred certification in 2003, Skagit was the largest low-impact hydroelectric project in the United States.

Source: LIHI, 2008a. Hydropower Reform Coalition, 2009.

- **Fuel cells.** Fuel cells combine oxygen and hydrogen to produce electricity without combustion, resulting in lower GHG emissions. However, fuel cells require a continuous stream of hydrogen-rich fuel and can only be considered a renewable energy technology if they operate on a renewably generated hydrogen fuel, such as digester gas or pure hydrogen generated by solar or wind energy generating systems (U.S. EPA, 2004).

RIALTO, CALIFORNIA – WASTEWATER TREATMENT
FACILITY FUEL CELL

Rialto, California installed a 900 kW fuel cell system
at its municipal wastewater treatment facility. The
facility collects used fats, oils, and grease from local
restaurants, and puts them through a digester to
produce methane. The fuel cell uses the methane to
produce electricity. The city expects that the system
will reduce electricity costs by $800,000 annually, and
will avoid nearly 5.5 million tons of CO_2 emissions. The
$15 million project cost will be partially lowered by a
$4 million rebate from the state, resulting in a payback
period of approximately 14 years.

Source: Chevron, 2007.

4. KEY PARTICIPANTS

A number of participants are may be helpful to planning and implementing on-site renewable energy generation projects at local government facilities and throughout the community, including:

- **Mayor or county executives.** The mayor or county executive can play a key role in increasing public awareness of the benefits of on-site renewable energy generation. Including on-site renewable energy generation in the mayor or county executive's priorities can lead to increased funding for renewable energy projects and broader implementation throughout the local government and the community.

The mayor of Newton, Massachusetts created the Mayor's Advisory Committee on Renewable Resources to facilitate PV system installations at the local high school, a community center, and in several residences (Newton, 2005).

In February 2007, the mayor of San Francisco announced an initiative to increase the city's solar power capacity from 2 MW to 35 MW. The mayor's plan encourages private and public entities to partner with the San Francisco Public Utilities Commission (PUC) to achieve this goal. In September 2007, as part of this initiative, the mayor unveiled a new PV system at the city's airport, a joint project between the airport and the PUC. The new system, which will generate 628,000 kWh of electricity annually, has helped move the city toward the mayor's goal of achieving 35 MW solar capacity. In addition, the project is expected to reduce CO2 emissions by 7,200 tons over the next 30 years (San Francisco, 2007; 2007a).

- **City and county councils.** Renewable energy generation activities are often initiated by the city and county council.

In Albuquerque, New Mexico, the city council passed a resolution that established a City Renewable Energy Initiative, which includes a requirement to retrofit all existing city-owned facilities with renewable energy generation systems. In addition, all new facilities over 100,000 square feet are to be equipped with renewable energy generation systems capable of producing enough energy to meet at least 25 percent of the facility's energy requirements (Albuquerque, 2005).

BAYONNE, NEW JERSEY SCHOOL DISTRICT –
PV INSTALLATION

In cooperation with the New Jersey Board of Public
Utilities, the Bayonne Board of Education committed
to installing nearly 10,000 solar panels at the local high
school and eight elementary schools. These panels have a
combined capacity of approximately 2 MW of energy, and
produce enough to power 200 small homes for 30 years.

The state's Clean Energy Program provided $5.4 million
worth of solar equipment and installation credits for the
$13.2 million project. In addition to reducing reliance on
fossil fuels, reducing pollution, and decreasing the local
strain on the electric grid, the project is expected to save
the school district more than $500,000 in avoided annual
electricity costs.

The solar power system at one of Bayonne's schools
played a role in maintaining power at the school, which
also serves as a community evacuation center, during
Superstorm Sandy in 2012. The solar panels provided
power to the school's emergency circuits, reducing the
load on diesel backup generators that otherwise could
have run out of fuel.

Source: New Jersey, 2006; Hunterdon County Democrat, 2012.

- **Planning departments.** Many local governments have modified local ordinances to facilitate on-site renewable energy generation system installation. These modifications often require coordination with local planning staff.

After the city's planning department rejected multiple proposed wind energy generation projects for not conforming to an existing ordinance, the city council of Fitchburg, Massachusetts asked the planning department to develop legislation that would modify the ordinance to include specifications for wind projects. In February 2008, the planning department presented the city council with proposed legislation that was approved following a public hearing (Butler, 2008).

- **School districts.** Installing renewable energy generation systems at schools can produce significant energy cost savings while also serving as an educational tool for demonstrating the benefits of renewable energy generation to students and the community at large. School boards can lead initiatives to install renewable energy generation systems at district facilities. For more information on how communities can reduce energy consumption in K-12 schools, see EPA's *Energy Efficiency in K-12 Schools* guide in the *Local Government Climate and Energy Strategy Series.*

EPA gave the school district of Spirit Lake, Iowa, a Green Power Partnership Award in 2012 for being among the nation's top 20 K-12 schools using renewable energy. The school district installed two wind turbines to power its elementary, middle, and high school, as well as the administrative buildings. Together the two turbines generate 1 MW of power, enough to supply 46 percent of the district's needs. The turbines saved the district more than $178,000 in energy costs in 2011 alone (U.S. DOE, 2012; Sioux City Journal, 2012).

- **Utilities.** Many utilities offer technical and financial assistance for on-site renewable energy generation system installation and operation. A number of local governments work with utilities to help local businesses and residents take advantage of these opportunities.

In Sacramento County, for example, six city councils have waived permitting fees for PV installations at the behest of the local municipal utility, which is providing additional incentives of $0.20 per watt AC to residential customers who install grid-connected PV systems (SMUD, 2007; DSIRE, 2013a).

Many local governments have worked with municipal electric utilities to adopt and implement net metering and renewable portfolio standard rules. Forty-three states and the District of Columbia have established net metering rules (DSIRE, 2013). These rules require utilities to allow customers to use excess energy generated on site to offset their consumption of energy from the grid. Where net metering rules are in place, affected utilities are required to measure the flow of electricity both to and from the customer. Customers pay for interconnection costs, but receive credit toward the following month's bill for net excess generation, typically at the utility's retail rate but sometimes at the lower wholesale rate, depending on state net metering rules.

Absence or presence of net metering rules can be an important consideration for local governments when planning on-site renewable energy generation systems. If there is a chance that a local government renewable energy generation system will produce more electricity than required, connection to the grid and net metering rules can ensure that the excess electricity is rewarded (U.S. DOE, 2006b; IREC, 2007).

The municipal electric utility in Tallahassee, Florida offers net metering for commercial and residential PV systems up to 10 kW in capacity (Tallahassee, 2007).

In March 2007, the city of New Orleans adopted net metering rules for utilities under its jurisdiction that mirrored the statewide net metering rules established in 2005 by the state Public Service Commission (DSIRE, 2007d).

- **Non-profit organizations.** Local governments can obtain technical and financial assistance from non-profit organizations to purchase and install on-site renewable energy generation systems.

Ashland, Oregon, for example, received a grant from a non-profit organization to establish the Ashland Solar Pioneer Program. The program installed solar PV systems at four locations, including two city-owned buildings. Excess energy generated from the PV systems is sold by the Ashland municipal utility to local customers (Ashland, 2007).

- **Energy service companies (ESCOs).** Many local governments have partnered with ESCOs to install renewable energy generation equipment on site at no upfront cost. Using performance contracting agreements, local governments can pay for installed equipment over time using energy cost savings. See Section 7, *Costs and Funding Opportunities*, for more information on ESCOs and performance contracting.

Cathedral City, California, partnered with an ESCO to install a $2.7 million PV system canopy at a city-owned parking garage. The ESCO installed the PV system at no upfront cost to the local government, and the city will pay for it from the energy cost savings. The system has already reduced the city's purchased power needs by 10 percent (Honeywell, 2006; American City and County, 2013).

- **Developers and financiers.** A number of local governments have purchased and installed PV systems through developers and financiers in an arrangement called the "solar services model" that partners local governments with developers who secure financing from a third party and install PV systems at local government facilities. For more information on the solar services model, see Section 7, *Costs and Funding Opportunities*.

In 2011, the San Jose Unified School District installed 5.5 MW of solar PV at 14 school district locations, including four high schools. This project was made possible by a partnership with Bank of America, which financed and owns the PV installations, and Chevron Energy Solutions, which installed and now maintains the PV systems. The district mounted the solar installations on parking lot shade structures and rooftops. The school district has integrated renewable energy and sustainability studies into its educational curriculum, and uses all 14 solar power

systems for educational purposes. The systems are expected to save the school $25 million over their lifetime and reduce the district's energy costs by 30 percent (NREL, 2011a).

5. FOUNDATIONS FOR PROJECT DEVELOPMENT

Local governments have used several mechanisms to initiate on-site renewable energy generation projects at their facilities and to adopt incentives for local businesses and residents, including:

- **Local government resolutions.** In some local governments, city and county councils must approve major alterations to government buildings and significant expenditures that require financing.

The city council in Ann Arbor, Michigan, passed a resolution in 2006 setting a goal for 5,000 solar hot water and PV systems to be installed across the city by 2015 (Ann Arbor, 2008; U.S. DOE, 2011). The city has installed a PV system on a science center and the city's farmer's market, and is working with more than 15 partners on outreach, education, and financing to promote additional installations.

- **Building energy codes.** Many local governments have adopted building energy codes, some of which include requirements that new buildings be designed to maximize potential for on-site renewable energy generation.

In Marin County, California, for example, the local building energy code requires new subdivisions to be designed to accommodate passive solar heating and cooling. Under the code, streets, lots, and building setbacks are to be arranged so that buildings are oriented with the long axis running east-west to maximize sunlight on the rooftop (Marin, 2008).

- **Net metering rules.** In states that do not have statewide net metering rules, or in states where net metering rules apply only to investor-owned utilities, local governments may be able to establish their own net metering rules for their own municipal utilities.

MODEL ORDINANCES FOR WIND TURBINES

A number of state governments, such as Massachusetts, have developed model zoning ordinances to facilitate local government siting of wind turbines. In other states, local governments have worked together to develop model zoning ordinances. In Minnesota, for example, several counties combined efforts with a non-profit organization and a regional development corporation to produce a *Model Wind Ordinance* and *Companion Document* in 2006.

Sources: MDER, 2007; MN Project, 2007.

CALIFORNIA LAW REQUIRES LOCAL GOVERNMENTS TO PERMIT CERTAIN WIND ENERGY PROJECTS

California passed legislation that prohibits local ordinances that unnecessarily impede the permitting of small wind projects. The legislation prohibits local governments from adopting ordinances that are more restrictive than the standards set forth in the legislation. The law effectively requires local governments to permit projects that meet these standards.

Source: California Assembly, 2001.

The city council of Yellow Springs, Ohio passed an ordinance requiring the municipal utility to provide net metering for customers, since the state's net metering rules apply only to investor-owned utilities (DSIRE, 2007c).

- **Renewable portfolio standards.** Many states have established renewable portfolio standards for investor-owned utilities or load-serving entities. These rules require utilities to meet a certain percentage of their energy supply with energy from qualified renewable sources. Some local governments have adopted similar requirements for municipal utilities.

In 2007, the city council of Austin, Texas passed a resolution that requires the municipal utility to use 30 percent renewable energy. The resolution requires that 100 MW of solar PV be used to generate the electricity to meet the 30 percent mandate (Austin, 2007).

- **Zoning ordinances.** Some local governments have found that modifications to zoning ordinances can facilitate renewable energy generation projects. For example, some zoning ordinances prohibit erection of structures that are in excess of 35 feet, a restriction that precludes installation of most wind turbines and some solar panels (U.S. DOE, 2005). A market survey of the "small wind" manufacturing industry identified restrictive zoning and permitting rules as the second most significant barrier to market expansion (after cost premiums) (Stimmel, 2007).

A number of local governments have adopted ordinances with specifications for wind turbines that have clarified and streamlined the local permitting process.

Rockingham County, Virginia, approved a zoning ordinance in 2004 that established specifications for permitting the installation of small wind turbines, including maximum turbine height, minimum parcel size, minimum setbacks, and noise limits (DSIRE, 2007b).

Mason City, Iowa, amended its existing zoning ordinance to allow for wind turbines to be installed in any zoning district (rather than in commercial or industrial zones exclusively) and to establish rules for siting turbines of 100 kW or greater (Mason City, 2006).

- **Ballot initiatives.** In some communities, constituent approval may be necessary to obtain funds for on-site renewable energy generation projects.

Columbia, Missouri, required voter approval to establish a local renewable portfolio standard for the city's municipal utility, with a goal to reach 2 percent generation from eligible renewable energy by 2008, increasing to 15 percent in 2023. By 2013, the city was generating 7.9 percent of its electricity from renewable energy sources, surpassing its interim goal by 2.9 percent (DSIRE, 2013b; U.S. DOE, 2012a; Columbia Water & Light, 2013).

Incentives for on-site renewable energy generation. Many local governments have established incentives for commercial and residential renewable energy generation projects. These incentives include:

> *Rebates.* Local governments have established financial incentives for residents and businesses to install renewable energy generation equipment.

In 2010, NYSERDA implemented a rebate program for eligible residential and nonresidential PV systems installations valued at $0.90–1.30/watt installed (NYSERDA, 2013).

> *Expedited permitting.* A number of local governments are facilitating commercial and residential on-site renewable energy projects for residents by expediting permitting processes.

Pike County, Illinois, has approved an ordinance that establishes specific criteria for wind energy projects. Providing developers with an explicit list of criteria for approval will help reduce the cost of designing on-site renewable energy generation systems (DSIRE, 2007a).

> *Permit credits and waivers.* A number of local governments have adopted permit credits or permit fee waivers to reduce the cost of installing on-site renewable energy generation systems.

In Tucson, Arizona, a city council resolution directed the Department of Development Services to offer building permit credits of up to $1,000 to applicants who install new PV, solar hot water and space heating, or solar air conditioning systems capable of producing a minimum of 1,500 kWh annually (Tucson, 2005).

In 2007, the San Bernardino County, California Board of Supervisors waived permit fees for installations of solar or wind power generation systems, solar hot water heaters, and energy-efficient heating and cooling systems on rooftops throughout the county. The fees for these permits had ranged from around $80 for water heaters to nearly $250 for wind turbines (U.S. DOE, 2007a; Gang, 2007).

> *Property tax credits and exemptions.* Some local governments have passed resolutions that modify local tax codes to provide incentives for local businesses and residents to install on-site renewable energy generation systems.

Harford County, Maryland, passed a resolution modifying the local tax code to offer property tax credits for facilities that use solar or geothermal systems. The credit is equal to the lesser value of one year of property taxes, or $2,500 per system or $5,000 per property per fiscal year. (DSIRE, 2013c).

6. STRATEGIES FOR EFFECTIVE PROJECT IMPLEMENTATION

Local governments have used a number of approaches to enhance the effectiveness of on-site renewable energy generation at their own facilities and throughout the community.

Bundle on-site renewable energy generation with energy efficiency improvements. Combining renewable energy generation with energy efficiency improvements that reduce energy loads enables local governments to meet a greater percentage of their electricity with electricity from renewable sources. In addition, the energy cost savings produced by energy efficiency improvements can be used to offset the purchase and installation costs of renewable energy generation systems and thus shorten payback periods.

Combine on-site renewable energy generation with green power purchases. Local governments can increase their GHG emissions reductions by combining on-site renewable energy generation with green power purchases.

Santa Monica, California, for example, has installed PV systems at multiple city facilities and is purchasing solar energy from an energy service provider to meet the balance of its energy needs (U.S. EPA, 2004; Santa Monica, 2007).

For more information on how local governments can implement green power purchases for their facilities and throughout the community, see EPA's *Green Power Procurement* guide in the *Local Government Climate and Energy Strategy Series*.

- **Coordinate with neighboring local governments.** By coordinating with other communities, local governments can achieve greater regional energy, environmental, and economic benefits. Encouraging on-site renewable energy generation throughout a region can lead to increased regional employment, reduced risk of energy supply disruption, and lower upfront costs due to market and technology maturation.

In Arizona, the Maricopa Association of Governments, representing a collection of communities around Phoenix, issued regional standard procedures for permitting PV system installations. Cities outside the association have since adopted the standards (Maricopa Association of Governments, 2002).

The Alaska Village Electric Cooperative, a member-owned utility cooperative that serves 55 villages in Alaska, has helped its member villages reduce their dependence on expensive fossil fuels by installing nearly 3,400 kW of wind power generating capacity. As of 2012, 12 communities had installed wind turbines (AVEC, 2012).

- **Engage the public.** Engaging businesses and residents in local government decision making can lead to enhanced support for on-site renewable energy generation projects. This support can be especially important given the significant local tax dollar investments required by many of these projects.

The Portsmouth, Rhode Island, Sustainable Energy Subcommittee conducted multiple public workshops to inform local residents and businesses of the town's efforts to construct wind turbines at two local schools. These workshops provided town staff the opportunity to address community questions and concerns (Portsmouth, 2007).

- **Evaluate energy generation capacity.** Because some renewable energy generation technologies have higher generation capacities in certain regions (e.g., wind power and solar PV), many local governments have conducted thorough evaluations of renewable energy generation potential for their facilities.

- **Sell renewable energy certificates (RECs).** RECs refer to the environmental attributes associated with the generation of renewable energy. These attributes can be separated from the renewable energy, allowing renewable energy generators to sell RECs on the market as a distinct product. The separated electricity, without its attributes, is then environmentally equivalent to conventional (i.e., non-renewable) electricity. RECs can be bought by organizations that do not have direct grid access to utility-provided green power, or do not have access to enough utility-provided green power to meet organizational goals (U.S. EPA, 2006b).

Local governments do not typically sell the energy they generate. However, local governments can take advantage of the market for RECs by selling the environmental attributes associated with the renewable energy they generate.

Local governments that sell their RECs can still benefit from stable, predicable electricity costs, but environmental claims are no longer valid. Because of the wide range of prices for RECs on the market, some local governments have been able to sell RECs from the electricity they generate while maintaining environmental claims by using revenues from REC sales to purchase lower-price RECs. Remaining revenues can be used to offset purchase and installation costs for renewable energy generation systems or to invest in other clean energy activities. Alternatively, local governments could sell RECs only for a period of time (e.g., until generation system purchase costs are recovered) and then retain the RECs to achieve the environmental and technological attributes (NREL, 2007b). (For more information on RECs, see EPA's *Green Power Procurement* guide in the *Local Government Climate and Energy Strategy Series*.)

7. COSTS AND FUNDING OPPORTUNITIES

This section provides information on the costs associated with on-site renewable energy generation, as well as information on how local governments can use multiple funding opportunities to address these costs.

Costs

Despite annual trends of declining costs, the cost premium associated with renewable energy generation systems can be significant (U.S. EPA, 2006b). Table 1, *Comparison of On-Site Renewable Energy Technology Costs*, provides rule-of-thumb approximations for costs associated with five renewable energy generation technologies and provides comparisons with the costs of other distributed generation systems that use conventional energy sources.

The installed cost of on-site renewable energy generation systems can be influenced by a range of factors, including the quality of the renewable resource in a given area, proximity of equipment manufacturers, and whether the installation was coupled with energy efficiency upgrades. While costs remain significant, they are decreasing. For example, according to a study by DOE, the cost of purchasing and installing wind turbines has decreased from approximately $3,500 per kW in 1985 to less than $1,300 per kW in 2005 (U.S. DOE, 2007). In addition, payback periods for on-site renewable energy generation systems are likely to continue to decrease electricity costs continue to rise. Availability of federal, state, local, and utility tax credits and rebates can also substantially reduce the payback period for these systems.

Funding Opportunities

Funding for local on-site renewable energy generation projects can come from a variety of sources, including:

- **Solar services model.**[5] Local governments have found that they can finance solar PV system purchases and installations at no upfront cost using the solar services model. Under this model, the local government signs a long-term (often 10 years) power purchase agreement with a developer and agrees to host a PV system at its

---

[5] *The solar services model is sometimes referred to as an independent energy purchase*

TABLE 1.   COMPARISON OF ON-SITE ENERGY GENERATION TECHNOLOGY COSTS

	Renewable					Conventional	
	On-Shore Wind Turbine	Solar PV	Geothermal	Fuel Cell	Biomass	Microturbine	Reciprocating Engine
Typical Project Size	5 kW–100 kW	10 kW–100 kW	2 tons–10 tons [k]	200 kW	5 MW–50 MW	25 kW–100 kW	5 kW–7 MW
Typical Total Installed Cost	$1,200–$2,100 per kW capacity [a]	$2,000–$6,800 per kW capacity [a]	$1,78–$9,900 per ton capacity [a,c]	$3,000–$4,000 per kW capacity [d]	$430–$4,200 per kW capacity [a]	$700 to $1,100 per kW capacity [e]	$1,075 per kW capacity [f]
Annual Operations and Maintenance Costs (fixed)	$11.70–$60 ($/kW-yr) [a]	$13–$110 ($/kW-yr) [a,b]	$150–$222 ($/kW-yr) [a]	$0.005–$0.010 per kW capacity [d]	$12–$87 ($/kW-yr) [a]	$0.005–$0.016 per kW capacity [e]	$0.005–$0.015 per kW capacity [f]
Life Span	30 years	30 years	30 years–45 years [h]	5 years–10 years [j]	30 years [i]	45,000 hours (~ 5 years) [g]	20 years–25 years

[a] NREL, 2012a.
[b] Lazard, 2012.
[c] Black & Veatch, 2012.

[d] WBDG, 2007.
[e] CEC, 2007.
[f] CEC, 2007a.

[g] U.S. DOE, 2003.
[h] REPP-CREST, Undated.
[i] EIA, 2003.

[j] WBDG, 2007.
[k] Geothermal unit capacity is measured in tons. One ton is equal to 12,000 Btu of energy per hour.

facility. The developer pays for the design, construction, and installation of the system, often arranging for third-party financing through an investor. The investor, who provides the upfront capital and owns the project, receives returns from payments from the host through the developer. The host's payments are at a pre-determined fixed price and are assessed much like a monthly utility payment. The local government, as host, benefits from fixed-price payments, reduced peak energy costs, and reduced GHG emissions, all at no upfront cost.

- **Lease-purchase agreements.** A tax-exempt lease-purchase agreement (also known as a municipal lease) allows public entities to finance purchases and installation over long-term periods using operating budget dollars rather than capital budget dollars. Lease-purchase agreements typically include "non-appropriation" language that limits obligations to the current operating budget period. If a local government decides not to appropriate funds for any year throughout the term, the equipment is returned to the lessor and the agreement is terminated. Because of this non-appropriation language, lease-purchase agreements typically do not constitute debt.

Under this type of agreement, a local government makes monthly payments to a lessor (often a financial institution) and assumes ownership of the equipment at the end of the lease term, which commonly extends no further than the expected life of the equipment. These payments, which are often less than or equal to the anticipated savings produced by the energy efficiency improvements, include added interest. The interest rates that a local government pays under these agreements are typically lower than the rates under a common lease agreement, because a public entity's payments on interest are exempt from federal income tax—meaning the lessor can offer reduced rates (U.S. EPA, 2004a).

In Hayward, California, a city council resolution authorized the city to install a solar power generating system at a local government facility using a 25-year lease purchase agreement. This agreement enabled the city to install the system at an annual lease payment of $70,400 (Hayward, 2005).

Unlike bonds, initiating a tax-exempt lease-purchase agreement does not require voter referendum to approve debt, a process that can delay renewable energy generation system installations. Tax-exempt lease-purchase agreements typically require only internal approval and an attorney's letter, a process that often takes only one week (as opposed to months or years for bonds). Local governments can expedite the process by adding renewable energy generation projects to existing master lease-purchase agreements (U.S. EPA, 2004a).

- **Performance contracting.** An energy performance contract is an arrangement with an ESCO that bundles together various elements of an energy-efficiency investment, such as installation, maintenance, and monitoring of energy-efficient equipment. These contracts, which often include a performance guarantee to ensure the investment's success, are typically financed with money saved through reduced utility costs but may also be financed using tax-exempt lease-purchase agreements (U.S. EPA, 2003).

CITY OF BOULDER, COLORADO – PERFORMANCE CONTRACTING

In 2009, Boulder, Colorado, entered an Energy Performance Contract (EPC) with the Colorado Energy Office. Under this EPC, the city was able to improve the energy efficiency for 66 city buildings whose upgrades would not cost taxpayers any extra money and would be paid back from the energy and water cost savings. As of 2013, the city was able to reduce its CO_2 emissions by 18% and energy use by 5,740MWh.

As of 2013, Boulder was already nearing the completion of the EPC, which involved installing 336 kW of solar energy, upgrading 28 more city facilities, and making other energy efficiency upgrades to reduce CO_2 emissions by another 2,000 metric tons.

Source: City of Boulder, 2013

Tucson, Arizona, used a performance contract to install solar pool heaters and domestic hot water heating systems at five public swimming pools at no up-front cost. The city uses its energy cost savings to pay for the systems (Apollo Alliance, 2006; U.S. Conference of Mayors, 2007).

Tax-exempt lease-purchase agreements are sometimes used to underwrite energy performance contracts with ESCOs. While local governments can often obtain financing directly from an ESCO, many have found that the interest rates available through tax-exempt lease-purchase agreements are typically lower than the rates offered by an ESCO. Tax-exempt lease-purchase agreements can be especially effective when used to underwrite energy performance contracts that include guaranteed savings agreements, under which an ESCO agrees to reimburse any shortfalls in expected energy cost savings.

Under the solar services model, the host is not responsible for performing or paying for maintenance on the system. Instead, those services are arranged and paid for by the developer. Ownership of the system can be transferred to the host when the developer's or financier's costs are recovered. Local governments that have used this model to install renewable energy generation systems at their facilities include Bend, Oregon and San Diego, California (Sandia, 2007; WRI, 2007; Bend, 2007; San Diego, 2007).

- **Local bonds.** A number of local governments have used bonds to finance renewable energy generation projects.

In 1981, Oregon passed the Oregon Small-Scale Energy Loan Program, which uses bond sales to finance small-scale energy projects in the state. By 2012, the program approved 854 loan applications, totaling around $594 million in renewable energy funds (DSIRE, 2013d).

- **State government.** Some states offer financial incentives to local governments that invest in on-site renewable energy generation. For example, NYSERDA provides cash incentives to local governments, colleges, and farms to offset purchase and installation costs of small wind turbines. Local governments can be eligible for up to $144,000, depending on the turbine model and the tower height (NYSERDA, 2007). Some states also offer financial assistance for local government officials to receive training in on-site renewable energy generation technologies.

The State of Alaska created the Renewable Energy Grant Fund in 2008 to help utilities, local governments, tribes, and other organizations study the feasibility, design, permitting, and construction of renewable energy projects. The fund will distribute up to $50 million annually until the program expires in 2023 (DSIRE, 2013e).

The Maryland Energy Administration provides grants for commercial and residential solar PV and hot water systems. Grants for businesses provide up to $60 per kW of PV capacity installed and up to $20 per square foot of solar hot water capacity installed. Residential installations receive flat awards of $1,000 per project for PV and $500 per project for solar hot water (DSIRE, 2012a).

- **Federal government sources.** Local governments can obtain financial assistance for purchasing and installing renewable energy generation systems from a variety of federal government sources. The U.S. Department of Energy (DOE), for example, provides grants and other financial incentives to local governments.

Portland, Oregon received a $200,000 grant through DOE's Solar America Cities program to install solar PV systems at city facilities and to fund the city's *Solar Now!* initiative, which installs solar PV systems at local residences (Ryan, 2007).

Local governments and their residents and businesses can find information on federal grants from more than two dozen government agencies at *http://www.grants.gov/*.

- **Non-profit organizations.** Non-profit organizations, such as independent foundations, can be a source of funding for local government renewable energy initiatives. A number of investor-owned utilities have created independent foundations to support clean energy initiatives.

An investor-owned utility established the Illinois Clean Energy Community Foundation in 1999 to invest in clean energy development and land preservation projects. Since awarding its first set of grants in 2001, the foundation has issued more than 3,900 grants totaling nearly $191 million. Many of these grants have been for local wind and solar projects. For example, the foundation awarded a $15,000 grant to a consortium of six counties in western Illinois to conduct a wind resource assessment study (ICECF, 2013).

- **Utilities.** Local governments can sometimes obtain financial assistance from utilities, many of which offer rebates or other incentives for on-site renewable energy projects.

In 2006, the California Public Utilities Commission established the California Solar Initiative, which requires three major California utility companies to provide solar installation incentives for their customers, including local government entities. This program estimates that it will provide more than $3 billion in solar incentives, and has set a goal to reach 1,940 MW of solar capacity by 2017. As of 2013, it was on track to meet this goal in late 2013 or 2014 (California Public Utilities Commission 2013; Renewable Energy World 2013).

Starting in 2010, Lakeland Electric, the municipal utility in Lakeland, Florida, began offering free installations of solar water heating systems in local homes. Residents pay only for water used, which is provided at a solar energy rate that is lower than the local electricity rate. The municipal utility benefits from reduced peak energy demand and from the sale of the RECs associated with the production of the renewable energy (DSIRE, 2007).

- **Voluntary ratepayer funding.** Some local governments have obtained funding for renewable energy generation projects from local residents.

Ellensburg, Washington used a unique financing approach that partnered local electricity customers with the city to install a 36 kW PV system. The city offered to reduce customers' future electricity bills in compensation for financial contributions toward the initial purchase and installation costs of the PV system. For example, if a customer were to contribute a certain percentage of the total funds contributed by all customers, that customer would receive that same percentage of the project's total solar power production, in the form of a deduction on their electricity bill (Ellensburg, 2007).

8. FEDERAL, STATE, AND OTHER PROGRAM RESOURCES

Local governments can obtain technical assistance and information from a number of federal, state, and other programs.

Federal Programs

- **U.S. EPA Green Power Partnership.** EPA's Green Power Partnership is a voluntary program to support the market for green power products. Local governments that meet partnership requirements earn publicity and recognition and are ensured of the credibility of their green power purchases. In addition, partners can receive EPA expert advice on identifying green power products and purchasing strategies, along with tools and resources to calculate the environmental benefits of green power purchases. The annual percentage requirements to qualify as a partner are as follows: 2 percent green power for entities using over 100 million kWh, 3 percent for between 10 million kWh and 100 million kWh, 6 percent for between 1 million kWh and 10 million kWh, and 10 percent for less than 1 million kWh.

  Website: *http://www.epa.gov/greenpower/*

- **ENERGY STAR®.** EPA's ENERGY STAR program provides a number of energy efficiency tools and resources that local governments can use when developing and implementing programs to reduce energy consumption. The ENERGY STAR Purchasing and

Procurement program, for example, provides lists of energy-efficient products (including geothermal heat pumps) with performance specifications, product savings calculators for assessing the cost-effectiveness of purchasing these products, sample procurement language, product retailer locators, and case studies.

Websites:

http://www.energystar.gov/

http://www.energystar.gov/index.cfm?c = geo_heat.pr_geo_heat_pumps

- **National Renewable Energy Laboratory (NREL).** NREL is the primary national laboratory for renewable energy and energy efficiency research and development. NREL provides local governments with information on existing and emerging technologies, including how to plan, site, and finance projects using renewable energy sources. NREL also provides information on developing rules and regulations for net metering and renewable portfolio standards for municipal utilities.

Website: *http://www.nrel.gov/learning/re_basics.html*

- **U.S. EPA State and Local Climate and Energy Program.** This program helps state, local, and tribal governments achieve their climate change and clean energy goals by providing technical assistance, analytical tools, and outreach support. It includes two programs:

  › The *Local Climate and Energy Program* helps local and tribal governments meet multiple sustainability goals with cost–effective climate change mitigation and clean energy strategies. EPA provides local and tribal governments with peer exchange training opportunities along with planning, policy, technical, and analytical information that support reduction of GHG emissions.

  › The *State Climate and Energy Program* helps states develop policies and programs that can reduce GHG, lower energy costs, improve air quality and public health, and help achieve economic development goals. EPA provides states with and advises them on proven, cost–effective best practices, peer exchange opportunities, and analytical tools.

Website: *http://www.epa.gov/statelocalclimate/*

- **U.S. Department of Energy Office of Energy Efficiency and Renewable Energy.** This office administers several programs that provide information and assistance for on-site renewable energy generation projects, including:

  › *Wind and Water Technologies Office.* Through this office, DOE works to improve wind energy technology development and deployment to help make wind energy competitive, and to develop new, cost-effective hydropower, marine, and hydrokinetic technologies that will have enhanced environmental performance and energy efficiency. *http://www1.eere.energy.gov/windandhydro/*

  › *Solar Energy Technologies Office.* Through this office, DOE partners state and local governments with national laboratories, universities, industry, and professional organizations to develop and deploy cost-effective technologies to expand the use of solar energy. *http://www1.eere.energy.gov/solar/*

  › *Geothermal Technologies Office.* DOE administers this office in partnership with the geothermal industry to establish geothermal energy as an economically competitive contributor to the U.S. energy supply. *http://www1.eere.energy.gov/geothermal/*

  › *Bioenergy Technologies Office.* Through this office, DOE provides information on biomass applications and potential and provides funding to develop the nation's biomass resources into cost-competitive, high-performing biofuels, bioproducts, and biopower. *http://www1.eere.energy.gov/bioenergy/*

  › *Fuel Cell Technologies Program.* Through this program, DOE works to develop hydrogen, fuel cell, and infrastructure technologies and to successfully introduce them in the mainstream market. *http://www1.eere.energy.gov/hydrogenandfuelcells/*

Website: *http://energy.gov/eere/office-energy-efficiency-renewable-energy*

State Programs

Many local governments work with state agencies to obtain technical assistance and information on purchasing and installing on-site renewable energy generating systems. Some states assist local governments in developing siting guidelines for on-site renewable energy system installations.

Michigan has developed guidelines for local governments to assist them in creating their own permitting requirements for wind energy projects. The guidelines include recommended zoning language for local governments to incorporate in ordinance modifications (Michigan DLEG, 2007).

Other Programs

- **American Council on Renewable Energy (ACORE).** ACORE, a non-profit organization composed of members from renewable energy industries, trade associations, financial institutions, governments, end-users, and other affiliated non-profits, promotes activities that support renewable energy technologies. Through its Renewable Energy Finance Network, the organization provides information on funding sources for renewable energy and energy efficiency projects.

  Website: *http://www.acore.org/*

- **American Wind Energy Association (AWEA).** AWEA is the U.S. trade association for wind power. The association includes more than 1,000 member organizations in the wind industry. AWEA promotes wind power as a clean, renewable resource through its annual industry conference, market reports and other research, education, and legislative efforts.

  Website: *http://www.awea.org/*

- **American Solar Energy Society (ASES).** ASES is a U.S. association for the solar energy industry. ASES publishes a monthly magazine, hosts the annual National Solar Tour and Solar Conference, organizes various solar industry conferences, and produces solar reports.

  Website: *http://www.ases.org/*

- **Geothermal Energy Association (GEA).** GEA is a U.S. association for the geothermal energy industry. GEA advocates for geothermal energy through an industry forum, research reports, annual conferences, and legislative efforts.

  Website: *http://www.geo-energy.org/*

- **National Hydropower Association (NHA).** NHA is the national association for the U.S. hydropower industry. The NHA is composed of almost 200 members from the hydropower industry and promotes hydropower through advocacy, legislative efforts, and educational outreach.

  Website: *http://www.hydro.org/*

- **Database of State Incentives for Renewables & Efficiency (DSIRE).** A project of the U.S. Department of Energy, the North Carolina Solar Center, and the Interstate Renewable Energy Council, DSIRE provides information on federal, state, and local incentives for renewable energy and energy efficiency projects, including tax credits, loans, and grants. The database also provides information on state and local regulations pertaining to renewable energy purchases and on-site renewable energy generation, including overviews of state and local net metering rules, renewable portfolio standards, and requirements for renewable energy use at public facilities.

  Website: *http://www.dsireusa.org/*

- **Interstate Renewable Energy Council (IREC).** IREC promotes deployment of renewable energy generation technologies at the state and local level by providing information and assistance to state and local governments for a number of renewable energy activities, including public education, procurement coordination, and adoption of uniform standards.

  Website: *http://www.irecusa.org/*

9. CASE STUDIES

The following case studies describe two comprehensive programs for generating renewable energy at local government facilities and reaching out to the community to involve local businesses and residents. Each case study describes how the program was initiated, key program features, and program benefits.

Boston, Massachusetts

PROGRAM INITIATION

Boston joined the ICLEI Cities for Climate Protection initiative in 2000 and hired an energy manager in 2001.

In 2005, the city's mayor signed the U.S. Mayors Climate Protection Agreement committing the city to meet Kyoto Protocol GHG emission reduction targets of 7 percent below 1990 levels by 2012, and 80 percent below 1990 levels by 2050. An executive order in 2007 reinforced this commitment, and required that all local government properties be evaluated for renewable energy generation potential. In 2007, the city also created the Boston Energy Alliance, a non-profit corporation to promote clean energy throughout the local government and the community. The city expects that the corporation's activities will involve participation of 25 to 35 percent of the city's electricity customers by 2013. Also in 2007, the U.S. Department of Energy named Boston as one of its Solar America Cities, leading to the establishment of the Solar Boston initiative.

PROFILE: BOSTON, MASSACHUSETTS

Area: 48 square miles

Population: 636,000 (2012)

Structure: Boston residents elect a mayor every four years, and the mayor can serve for multiple terms. Members of the Boston City Council are elected every two years. The city's renewable energy activities, including the Solar Boston initiative, are directed by the Department of the Environment.

Program Scope: Boston has installed wind and PV energy systems at a broad range of facilities, including government buildings, public schools, and affordable housing units, and has worked with the private sector to encourage renewable energy generation at a number of businesses and residences.

Program Creation: Boston launched the Solar Boston initiative following the city's selection as a DOE Solar America City in 2007. Previous activities, including participation in the ICLEI Cities for Climate Protection and the U.S. Mayors Climate Protection Agreement, contributed to the development of this initiative.

Program Results: By the end of 2008, the city expected to have a combined total of 1 MW solar PV capacity installed. Recent solar PV mapping efforts reveal that the city could meet 14 to 19 percent of its electricity supply using PV systems.

PROGRAM FEATURES

- **Combining on-site renewable energy generation with green power purchases.** In addition to pursuing wind, solar, and biomass energy options, 11 percent of the electricity the city purchases comes from green power sources. This commitment will increase to 15 percent by 2012, as directed by a 2007 executive order (Boston, 2007).

- **Solar America City.** DOE has named Boston as one of its "Solar America Cities." The Solar America Cities partnership is an initiative to establish model local governments that help improve national solar infrastructure and facilitate mainstream adoption of solar technologies (Boston, 2008a).

- **Solar mapping.** As part of its activities under the Solar Boston initiative, Boston has used GIS technology to evaluate and map the potential for PV systems throughout the city. The mapping revealed a total potential capacity of between 670 MW and 900 MW on rooftops across the city. A similar effort is underway to evaluate the potential for solar water heating applications (IREC, 2008).

- **Affordable housing.** The city is working with the Department of Neighborhood Development and the Boston Housing Authority to encourage installing solar PV equipment on affordable housing units. The Boston Housing Authority is investigating options for using a performance contract to install between 115 kW and 120 kW of PV capacity (Boston, 2008a). (For information on how Boston is improving the energy efficiency of its affordable housing, see EPA's *Energy Efficiency in Affordable Housing* guide in the *Local Government Climate and Energy Strategy Guides* series.)

- **Boston Energy Alliance.** This non-profit corporation was formed to facilitate the city's energy efficiency and renewable energy activities. The corporation will use a $300 to $500 million revolving loan fund to finance energy efficiency improvements and renewable energy generation systems at city facilities (Boston, 2008a).

- **Solar workshop.** In January 2008, the city hosted a workshop to present its goals for the future of solar in the city, and to invite stakeholders to participate in city activities (Boston, 2008c).

- **Wind power.** The city is working with the Massachusetts Technology Collaborative (MTC) to assess the feasibility of installing wind turbines on an island in the city's harbor. This study is being conducted in conjunction with a study by the state Water Resources Authority to install another turbine in the harbor.

The city is also supporting an initiative by the Community Wind Collaborative to install small wind turbines throughout the city using $4 million in funds from the MTC Renewable Energy Trust. In addition, the city plans to install a 1.8 kW turbine at the city hall in 2008 and is in the process of developing a Wind Energy Zoning provision for the local zoning code that would streamline siting of wind turbines in the city (Boston, 2008; Boston, 2008b; Boston, 2008d).

PROGRAM RESULTS

The city's initiative has resulted in solar PV installations on a broad range of buildings, including government buildings, housing developments, and public schools (Boston, 2007a). Under the Solar Boston initiative, the city expected to reach nearly 1 MW of total installed PV capacity in 2008 (1.8 MW if PV installations on affordable housing units are included) (Boston, 2008a). The solar PV mapping activities, which revealed a total capacity ranging between 670 MW and 900 MW, indicate that solar PV could supply between 14 and 19 percent of the entire city's electricity demand (IREC, 2008).

Website: *http://www.cityofboston.gov/environmental-andenergy/conservation/solar.asp*

Waverly, Iowa

Waverly Light and Power, a municipal electric utility owned by the city of Waverly, Iowa, was the first municipal utility in the country to generate its own wind power. The utility has set a goal of meeting 20 percent of its energy needs by 2020 with its own on-site renewable energy.

PROGRAM INITIATION

The utility's energy demand grew dramatically in the 1980s, leading city planners to consider alternative energy supplies. When the utility's purchased power contract terminated in 1991, it conducted a study to assess the feasibility of generating energy from renewable sources. In 1993, with grants from the American Public Power Association, Waverly Light and Power installed its first wind turbine (ICLEI, Undated; U.S. DOE, 2003).

PROFILE: WAVERLY, IOWA

Area: 33 square miles

Population: 4,900 customers

Structure: Waverly Light and Power is a city-owned utility governed by the city-nominated Waverly Light and Power Board and a board-selected general manager.

Program Scope: Waverly Light and Power's Green Choice and Energy Tags programs are available to the municipal utility's electric customers.

Program Creation: Waverly Light and Power installed its first wind turbine in 1993 in an effort to reduce the utility's rising energy costs.

Program Benefits: The utility is obtaining 5 percent of its electricity from renewable sources. This effort has resulted in CO2 emission reductions as high as 7,000 tons in one year.

Source: ICLEI, Undated; Waverly, 2007.

PROGRAM FEATURES

To complement its first turbine purchased in 1993 (an 80 kW system), Waverly Light and Power purchased two additional 750 kW turbines in 1999. In 2002, it replaced the first turbine with a 900 kW turbine, which produces nearly 1.85 million kWh annually. In 2005, in order to purchase additional land on which new, state-of-the-art turbines could be built, the utility sold its two 750 kW turbines. With the addition of two new 900 kW turbines purchased in 2007, the utility planned to produce nearly 6 million kWh annually beginning in 2008.

The cost of the turbines was financed in part by grants (from the American Public Power Association and from NREL) and in part from the utility's capital budget. The currently active turbine cost just over $1 million installed. Maintenance costs typically reach approximately $1,500 per year per turbine. The utility offers its customers the opportunity to purchase some of this green power through its Green Choice program. Many local customers pay a premium of less than $2 per month to receive green power.

In addition to generating and selling renewable energy, Waverly Light and Power became the first U.S. utility to offer RECs in 2001. Under its Iowa Energy Tags Program, Waverly Light and Power sells the environmental attributes of the green power it produces to help pay for future investments in wind energy. Each tag represents 2,500 kWh of green power, which translates into a savings of more than two tons of CO_2 emissions (WLP, 2007; WLP, 2007a; WLP, 2006; WLP, 2005; ICLEI, Undated; U.S. DOE, 2003).

PROGRAM BENEFITS

In 2002, Waverly Light and Power's wind turbines reduced the city's CO_2 emissions by nearly 7,000 tons. In 2005, the CO_2 emissions reduction was approximately 4,300 tons (following the sale of the two 750 kW turbines to raise money for investments in newer wind technologies). In addition, the sale of 800 energy tags earns the utility approximately $40,000 annually, money that is earmarked for investment in new renewable energy sources. Through 2003, Waverly Light and Power was meeting 5 percent of its energy demands from wind power (ICLEI, Undated; WLP 2006).

Website: *http://wlp.waverlyia.com/renewable_energy.asp*

10. ADDITIONAL EXAMPLES AND INFORMATION RESOURCES

Title/Description	Website
Examples of On-Site Renewable Energy Generation	
Albuquerque, New Mexico. Albuquerque established a Renewable Energy Initiative that includes retrofitting existing public buildings with renewable energy generating systems. In addition, all new facilities over 100,000 square feet are required to meet 25 percent of their energy use with renewable energy.	*http://www.cabq.gov/energy/documents/Resolution329.doc*
Anaheim, California. The city installed a 12,500 square foot solar array on the roof of the Anaheim Convention Center, the largest convention center on the West Coast. This system generates 140,000 kWh of electricity annually.	*http://www.anaheim.net/utilities/adv_svc_prog/renew_energy/renew.html*
Ann Arbor, Michigan. The city council in Ann Arbor, a DOE-awarded Solar America City, passed a resolution that sets a goal of having 5,000 PV systems installed throughout the city by 2015. The city achieved its goal earlier than expected.	*http://www1.eere.energy.gov/solar/pdfs/50192_annarbor.pdf*
Ashland, Oregon. Ashland established a Solar Pioneers Program that involved the installation of solar energy systems at city buildings. Energy produced by these systems is delivered to residents who fund the program by voluntarily contributing to the municipal utility's solar surcharge. The program was such a success that Ashland initiated a Solar Pioneers II Program, which funded the installation of a 63 kW solar system on the City Service Center.	*http://www.ashland.or.us/Page.asp?NavID=14017*
Austin, Texas. Austin, a DOE-awarded Solar America City, established a renewable portfolio standard that requires the local municipal utility to deliver 35 percent renewable energy by 2020. The standard is required to be met, in part, by 200 MW of solar PV power.	*http://energy.gov/savings/austin-renewables-portfolio-standard*
Berkeley, California. The City of Berkeley installed a 1.8 kW wind turbine at its Shorebird Nature Center to produce electricity for saltwater aquariums, computers, and lighting. The turbine, which was specifically designed to generate electricity in low winds, is reducing the city building's annual GHG emissions by 80 percent.	*http://www.ci.berkeley.ca.us/citycouncil/2006citycouncil/packet/032106/2006-03-21%20Item%2013%20Wind%20Turbine%20at%20Shorebird%20Nature%20Center.pdf*
Boise, Idaho. Boise completed a geothermal loop that recycles geothermal water used to heat more than 65 buildings in the city. Based on the project's success, Boise developed a geothermal system to generate heat and power for the Boise State University campus.	*http://publicworks.cityofboise.org/services/geothermal/* *http://news.boisestate.edu/update/2012/11/16/geothermal-now-heating-up-campus/*

Title/Description	Website
Boston, Massachusetts. Boston installed a 1.9 kW wind turbine on the roof of the city hall. As of 2013, the city was studying the feasibility of more urban rooftop wind turbines to comply with the Massachusetts Green Communities Act, which requires the state to generate 20 percent of its electricity from renewable energy sources.	http://www.wpi.edu/Pubs/E-project/ Available/E-project-050410-163916/ unrestricted/mc_rd_an_js_Rooftop_ Wind_IQP_Report.pdf
Cayuga County, New York. Cayuga County constructed an anaerobic digestion facility to turn waste from local dairy farms and food processors into biofuels to produce electricity and heating for local county buildings. This system generates 5,157,000 kWh of electricity annually.	http://www.osti.gov/bridge/product. biblio.jsp?osti_id=1070030
Chico, California. Chico installed a combined 1.2 MW of PV power at two municipal facilities, the Water Pollution Control Plant and the downtown parking system. The PV systems provide 19 percent of the city's municipal load and reduce CO2 emissions by nearly 750 metric tons annually.	http://us.sunpowercorp.com/cs/Sate llite?blobcol=urldata&blobheader=a pplication%2Fpdf&blobheadername3 =Content-Disposition&blobheaderva lue3=attachment%3B+filename%3D9 04%252F662%252Fsp_chicocity_en_ ltr_p_cs.pdf&blobkey=id&blobtable= MungoBlobs&blobwhere=130025852 1092&ssbinary=true
Culver City, California. The city council of Culver, California, passed an ordinance requiring that all new constructions for commercial and residential buildings in the city must install 1 kW of PV energy generation for every 10,000 square feet. As part of this requirement, the city council waived the solar permit fees to reduce the cost of installations.	http://www.culvercity.org/~/media/ Files/BuildingSafety/AlternateEnergy/ MandatorySolarPhotovoltaic Requirement_Spring2008%20pdf.ashx
Hayward, California. The city installed a 1 MW solar system for its Water Pollution Control Facility. This system will generate an estimated 1.95 MW annually and contribute to 24 percent of the facility's waste treatment energy needs.	http://www.businesswire.com/news/ home/20110309006551/en/City- Hayward-REC-Solar-Unveil-1-MW
Kotzebue, Alaska. Kotzebue installed 17 wind turbines to provide electricity to its residents. This 1.14 MW system provides an alternative to electricity generation from diesel generators, which require fuel that must be shipped over 1,200 miles.	http://www.kea.coop/articles/the- wind-farm-that-continues-to-grow/
Mackinaw City, Michigan. Mackinaw City installed two wind turbines that combined produce approximately 2,000,000 kWh annually and are being used to teach students at the local community college.	http://www.mackinawcity.org/wind- turbine-generators-140/
Minneapolis, Minnesota. Minneapolis installed solar energy generating systems on four city buildings (including a convention center), and developed a partnership with the city of Saint Paul that led to a series of joint climate protection agreements and greenhouse gas reduction initiatives.	http://www1.eere.energy.gov/solar/ pdfs/51054_minneapolis_stpaul.pdf
Murray City, Utah. Murray City adopted rules that require the municipal utility to provide bi-directional net metering to customers who produce up to 10 kW of power from solar, wind, or hydroelectric sources.	http://www.dsireusa.org/library/ includes/incentivesearch. cfm?Incentive_Code = UT11R&Search = TableType&type = Net&CurrentPageID = 7&EE = 0&RE = 1
North Bonneville, Washington. North Bonneville installed a geothermal exchange system in its city hall. The system reduces energy costs by an estimated $1,500 annually.	https://www.geothermal-library.org/ index.php?mode=pubs&action=view &record=1015540
Portland, Oregon. The Portland Department of Transportation uses solar energy to power vehicles and parking meters. The Portland recycling facility runs on a 10 kW wind turbine. In addition, Portland developed a solar highway, composed of 594 solar panels along the side of Interstate 5 and Interstate 205 South. This 104 kW system contributes one-third of the energy needed to power lighting on the highway.	http://www.portlandoregon.gov/ transportation/47418 http://www.oregon.gov/ODOT/HWY/ OIPP/pages/inn_solarhighway.aspx
San Francisco, California. The San Francisco Public Utilities Commission installed 24,000 solar panels on the roof of Sunset Reservoir. This 5 MW system provides clean energy to the city's municipal facilities.	http://www.sfbuildingtradescouncil. org/content/view/461/65/

Title/Description	Website
Yarmouth, Maine. Students at Yarmouth High School initiated a solar power project involving the construction of a 3,600 kW system that offsets more than 1.8 metric tons of CO2 emissions annually.	*http://www.maine.gov/mpuc/ staying_informed/news/ PRVRRFYarmouth10-23-2007.doc*
Examples of Incentives for Residential or Commercial On-Site Renewable Energy Generation	
Honolulu, Hawaii. The County of Honolulu offers qualified homeowners zero-interest loans to install solar water heating and PV systems at their homes.	*http://www.dsireusa.org/incentives/ incentive.cfm?Incentive_Code=HI15F*
Huntington Beach, California. Huntington Beach has adopted an Energy-Efficient Permit Fee Waiver Program for solar equipment that produces renewable energy on-site, including PV systems and solar water heating systems.	*http://www.ci.huntington-beach. ca.us/files/users/planning/planning- newsletter-4q-2012.pdf*
San Bernardino, California. The San Bernardino Board of Supervisors approved a waiver for building permit fees for installation of solar energy systems, wind turbines, tankless water heaters, and energy-efficient HVAC systems.	*http://www.sbcounty.gov/ greencountysb/building_permit_fee_ waivers.aspx*
Santa Clara, California. The city of Santa Clara offers residents and businesses the opportunity to rent city-owned solar equipment. Residents pay a small installation fee in addition to the costs of energy from the grid, but are allowed to keep the energy cost savings from the on-site renewable energy generated.	*http://www.dsireusa.org/ incentives/incentive.cfm?Incentive_ Code=CA19F*
Thief River Falls, Minnesota. Thief River Falls offers local residents loans of up to $8,000 at 5 percent interest and a rebate of $2,000 for the installation of ground source heat pumps.	*http://www.citytrf.net/Printable_ forms.htm*
Examples of Ordinances Supporting Renewable Energy Generation	
Ashe County, North Carolina. In 2007, Ashe County passed a local ordinance that adopted regulations to streamline permitting of wind energy generation systems.	*http://www.dsireusa.org/ incentives/incentive.cfm?Incentive_ Code=NC11R*
Columbia, Missouri. In 2004, The City of Columbia approved a renewable energy ordinance that requires the municipal utility to purchase increasing levels of energy from renewable resources, rising to 15 percent of electric retail sales by 2022.	*http://www.gocolumbiamo.com/ WaterandLight/Documents/ RenewReport.pdf*
Maricopa County, Arizona. The Maricopa County Zoning Ordinance contains provisions for siting renewable energy systems, allowing renewables to be installed in any zoning district within the county as long as certain siting requirements are met.	*http://energy.gov/savings/maricopa- county-renewable-energy-systems- zoning-ordinance*
Information Resources on On-Site Renewable Energy Generation	
American Wind Energy Association. The American Wind Energy Association has a number of helpful resources available to consumers interested in on-site renewable energy, including fact sheets and cost estimates.	*http://www.awea.org/Resources/ index.aspx?navItemNumber=506*
2012 Wind Technologies Market Report. This DOE report provides statistics on national wind power capacity, turbine size and cost, wind power prices, and policy efforts driving wind power development.	*http://www1.eere.energy.gov/wind/ pdfs/2012_wind_technologies_ market_report.pdf*
APS Energy. APS Energy partnered with multiple local governments to install solar energy systems at municipal facilities. This APS Energy website provides information on a sample of these projects.	*http://www.aps.com/en/ communityandenvironment/ environment/solarinitiatives/Pages/ home.aspx*
Community Jobs in the Green Economy. This Apollo Alliance and Urban Habitat report describes the potential for job creation from investing in energy efficiency and renewable energy.	*http://www.urbanhabitat.org/files/ Community-Jobs-in-the-Green- Economy-web.pdf*
FEMP Renewable Energy. The DOE FEMP program provides information on federal government initiatives for using renewable energy, including on-site generation.	*https://www1.eere.energy.gov/femp/ technologies/renewable_energy.html*

Title/Description	Website
Fuel Cell Technology. This website, developed as a component of the Whole Building Design Guide, provides information on fuel cell technologies and applications.	*http://www.wbdg.org/resources/ fuelcell.php*
Geothermal Energy in County Facilities. The National Association of Counties developed this resource to provide information on the costs and benefits of applying geothermal technologies in local government facilities.	*http://www.naco.org/programs/ csd/Green%20Government%20 Documents/EE_Factsheet%20-%20 Geothermal%20Energy%20in%20 County%20Facilities.pdf*
Geothermal Heat Pumps. This DOE website provides information on the basics of geothermal exchange. The site includes fact sheets on the logistics of using geothermal heat pumps in different building types.	*http://energy.gov/energysaver/ articles/geothermal-heat-pumps*
Geothermal Resources Maps. DOE collected information on the location of geologic resources that could make geothermal applications potentially feasible.	*http://www1.eere.energy.gov/ geothermal/maps.html*
Government Facilities Case Studies. The Geothermal Heat Pump Consortium collected fact sheets on municipal government examples of geothermal applications.	*http://www.geoexchange.org/ federal/case.htm*
Guide to Purchasing Green Power. This EPA guide provides information on purchasing green power. Chapter 7 addresses on-site renewable energy projects.	*http://www.epa.gov/greenpower/ pdf/purchasing_guide_for_web.pdf*
High Performance Technologies: Solar Thermal & Photovoltaic Systems. This DOE report provides information on building zero-energy homes using solar thermal and PV technologies. Local governments can find information on site planning and orientation of solar thermal and PV applications.	*http://apps1.eere.energy.gov/ buildings/publications/pdfs/building_ america/41085.pdf*
Green Power from Landfill Gas. This EPA fact sheet provides information and statistics on how landfills can be used to produce electricity in a manner that is protective of natural resources.	*http://www.epa.gov/lmop/ documents/pdfs/LMOPGreenPower. pdf*
Jobs from Renewable Energy and Energy Efficiency. This fact sheet provides information on existing and projected energy efficiency- and renewable energy-related jobs in the U.S. by sector.	*http://www.eesi.org/fact-sheet- jobs-renewable-energy-and-energy- efficiency-01-jun-2011*
Money from the Sun: An Investor's Guide to Solar-Electric Profits. This article describes the long-term benefits of investing in solar energy systems.	*http://www.eerl.org/index. php?P=FullRecord&ID=3822*
Potential for Energy Efficiency, Demand Response, and On-site Renewable Energy to Meet Texas's Growing Electricity Needs. This ACEEE report provides policy recommendations to meet growing energy demand in Texas. Recommendations include development of the public buildings program and providing incentives for onsite renewable energy.	*http://aceee.org/research-report/ e073*
Putting Renewables to Work: How Many Jobs Can the Clean Energy Industry Generate. This University of California–Berkeley report shows the economic benefits of investing in renewable energy in terms of jobs created.	*http://rael.berkeley.edu/files/2004/ Kammen-Renewable-Jobs-2004.pdf*
Renewable Energy and Distributed Generation Guidebook. This Massachusetts Division of Energy Resources report provides an overview of implementation issues associated with siting and generating distributed energy and connecting to the grid.	*http://www.mass.gov/eea/docs/doer/ pub-info/guidebook.pdf*
Renewable Energy and Energy Efficiency: Economic Drivers for the 21st Century. This report was developed by the American Solar Energy Society to describe the existing and projected breakdown of renewable energy and energy efficiency-related employment in the United States.	*http://www.greenenergyohio.org/ page.cfm?pageID=2257*
Renewable Energy Basics. NREL provides basic information on seven forms of renewable energy applications.	*http://www.nrel.gov/learning/re_ basics.html*

Title/Description	Website
The Role of Distributed Generation in Power Quality and Reliability. This NYSERDA report assesses the power quality and supply reliability benefits of distributed generation technologies, including on-site renewable energy generation.	http://www.localpower.org/documents/report0_nyserda_reliability.pdf
Single, Paired, and Aggregated Anaerobic Digester Options. This study presents an overview of the feasibility of installing anaerobic digesters to turn dairy farm waste into usable biofuels.	http://www.manuremanagement.cornell.edu/Pages/General_Docs/Fact_Sheets/Perry_Feas_Study_factsheet.pdf
Solar America Initiative. In 2007, DOE launched the Solar America Initiative to accelerate solar power applications in 13 model cities across the United States.	http://www1.eere.energy.gov/solar/solar_america/index.html
Solar Energy Industry Association. This solar trade association provides resources and research on how to advance solar implementation in the United States.	http://www.seia.org/research-resources
Solar Swimming Pool Heaters. This fact sheet provides information on the technical aspects and benefits of installing solar water heaters on swimming pools.	http://energy.gov/energysaver/articles/solar-swimming-pool-heaters
Using Distributed Energy Resources. This DOE fact sheet provides an overview of the benefits of using distributed energy resources, such as on-site energy generating technologies, and describes the process for determining the need for distributed energy resources at a facility.	http://www1.eere.energy.gov/femp/pdfs/31570.pdf
Wind Energy Economics. The Iowa Energy Center developed this resource to provide information on the cost-effectiveness of wind turbines.	http://www.iowaenergycenter.org/wind-energy-manual/wind-energy-economics/
Resources on Financing for On-Site Renewable Energy Generation	
The Borrower's Guide to Financing Solar Energy Systems. This brochure provides an overview of financial assistance opportunities offered by the federal government and private lenders for the installation of on-site renewable energy systems.	http://www.nrel.gov/docs/fy99osti/26242.pdf
DSIRE. The Database of State Incentives for Renewable Energy provides information on state and local government renewable energy and energy efficiency incentives.	http://www.dsireusa.org/
Energy Tax Incentives. The Tax Incentives Assistance Project, a collaborative of non-profit organizations, government agencies, and other stakeholders, provides consumers and businesses with information on incentives available through the federal Energy Policy Act of 2005.	http://www.energytaxincentives.org/
Federal Grants. The federal grants.gov program provides information on financial incentives available from 26 government agencies for a range of investments, including renewable energy generation.	http://www.grants.gov/
Handbook on Renewable Energy Financing for Rural Colorado. This handbook provides information on state and federal resources for financing renewable energy projects in Colorado. Many of the resources identified may be relevant to local governments outside Colorado.	http://www.windpoweringamerica.gov/filter_detail.asp?itemid=1101
Innovations in Renewable Energy Financing. This National Renewable Energy Laboratory paper provides information on new strategies for financing renewable energy projects, including REC sales.	http://www.usaee.org/usaee2007/submissions/OnlineProceedings/Innovations%20in%20Renewable%20Energy%20Financing%20_Cory_%20-%20FINAL.pdf
Tools for Screening On-Site Renewable Energy Generation Applications	
Clean Power Estimator. This tool provides quick cost-benefit analysis for PV, solar thermal, wind, and energy efficiency technologies for residential and commercial buildings in specified geographic regions.	http://www.gosolarcalifornia.org/tools/clean_power_estimator.php

| --- | --- |
eGRID. EPA's eGRID is a comprehensive source of data on the environmental characteristics of domestic electric power generation. It compiles data from 24 federal sources on emissions and resource mixes for virtually every power plant and company that generates electricity in the United States. It also provides user search options, including aspects of individual power plants, generating companies, states, and regions of the power grid.	http://www.epa.gov/cleanenergy/egrid/index.htm
EPA Power Profiler Tool. This EPA tool provides emission factors for a given region to help calculate the pollution benefits of energy savings. Users enter a ZIP code and specify their electric utility. This tool uses information from EPA's eGRID database of emissions and electricity generation data.	http://www.epa.gov/cleanenergy/powerprofiler.htm
Find Solar. Find Solar is a collaborative project involving DOE and the American Solar Energy Society that enables a user to calculate the costs, savings, and GHG emissions reductions of converting a portion of a building's energy use to solar generation.	http://www.findsolar.com/index.php?page = rightforme
FRESA. The Federal Renewable Energy Screening Application was developed by DOE as a tool for assessing the comparative benefits of different renewable energy applications at federal facilities.	https://www3.eere.energy.gov/femp/fresa/
ProForm. This tool, developed by LBNL, calculates the financial indicators and reduced GHG emissions of renewable energy projects.	http://poet.lbl.gov/Proform/
PV Watts. This NREL performance calculator estimates the energy and cost savings from grid-connected PV systems from various locations around the country. The user can adjust various data assumptions to accommodate for regional and system specifics.	http://www.nrel.gov/rredc/pvwatts/
RETScreen. This international tool was developed by Natural Resources Canada to evaluate the energy production and savings, emissions reductions, and financial viability of different types of energy efficiency and renewable energy investments, including on-site renewable energy generation.	http://www.retscreen.net/ang/d_o_view.php

11. REFERENCES

Albuquerque. 2005. *Resolution R-05-329: Adopting Policies to Establish and Implement a City Renewable Energy Initiative.* Available: http://www.cabq.gov/energy/documents/Resolution329.doc. Accessed 7/17/2007.

American City and County. 2013. *Energy Plan Reaps Environmental and Economic Benefits.* Available: http://americancityandcounty.com/energy-efficiency/energy-plan-reaps-environmental-and-economic-benefits. Accessed 8/29/2013.

Anaheim. 2001. *Anaheim Public Utilities Advantage Services Incentive Programs Recognized by State Association.* Available: http://www.anaheim.net/utilities/news/article.asp?id = 245. Accessed 9/28/2007.

Ann Arbor. 2008. *5,000 Solar Roofs.* Available: http://www.a2gov.org/government/publicservices/systems_planning/energy/energychallenge/Pages/SolarRoofs.aspx. Accessed 3/12/2008.

Apollo Alliance. 2006. *New Energy for Cities.* Available: http://www.apolloalliance.org/docUploads/new_energy_cities.pdf. Accessed 7/17/2007.

Ashland. 2007. *Ashland Solar Pioneer Program.* Available: http://www.ashland.or.us/Page.asp?NavID = 1534. Accessed 7/18/2007.

Austin. 2007. *Austin Climate Protection Plan.* Available: http://www.ci.austin.tx.us/council/downloads/mw_acpp_points.pdf. Accessed 7/18/2007.

Austin Energy. 2007. *Rebates and Loans.* Available: http://www.austinenergy.com/Energy%20Efficiency/Programs/Green%20Building/Programs/rebates-Loans.htm. Accessed 7/26/2007.

AVEC. 2012. *AVEC Wind Program Recap.* Available: http://www.avec.org/renewable-energy-projects/Wind%20Program%20Recap.pdf. Accessed 8/30/2013.

AWEA. 2007. *Wind Energy Basics.* Available: http://www.awea.org/faq/wwt_basics.html. Accessed 9/27/2007.

AWEA. 2007a. *The Economics of Small Wind.* Available: http://www.awea.org/smallwind/toolbox2/factsheet_econ_of_smallwind.html. Accessed 7/9/2007.

Black & Veatch. 2012. *Cost & Performance Data for Power Generation.* Available: http://bv.com/docs/reports-studies/nrel-cost-report.pdf.

Bend. 2007. *Solar Services Agreement.* Available: http://www.ci.bend.or.us/city_hall/meeting_minutes/docs/IS_SunEnergy_Solar_Agreements.pdf. Accessed 1/31/2008.

Berkeley. 2007. *Services, Incentives, and Energy Rebate Programs.* Available: http://www.cityofberkeley.info/sustainable/residents/ResSidebar/EnergyServices.html. Accessed 7/26/2007.

Boston. 2007. *Executive Order: Relative to Climate Action in Boston.* Available: http://www.cityofboston.gov/environmentalandenergy/pdfs/Clim_Action_Exec_Or.pdf. Accessed 3/28/2008.

Boston. 2007a. *Sustainability Accomplishments.* Available: http://www.cityofboston.gov/environmentalandenergy/pdfs/sus_accom.pdf. Accessed 3/27/2008.

Boston. 2008. *Boston Plans Wind Turbines for City Hall, Schools.* Available: http://boston.about.com/od/governmentcityservices/a/WindTurbines.htm. Accessed 3/11/2008.

Boston. 2008a. *Thinking BIG About Boston's Solar Energy Future.* Available: http://www.cityofboston.gov/climate/pdfs/SolarBostonPresentation-1-10-08.pdf. Accessed 3/27/2008.

Boston. 2008b. *Wind Energy.* Available: http://www.cityofboston.gov/environmentalandenergy/wind.asp. Accessed 3/27/2008.

Boston. 2008c. *Solar Boston.* Available: http://www.cityofboston.gov/climate/solar.asp. Accessed 3/27/2008.

Boston. 2008d. *Mayor Announces Plans for Wind Turbines.* Available: http://www.cityofboston.gov/climate/solar.asp. Accessed 3/27/2008.

Boulder County. 2013. *Sustainable Energy Plan.* Available: http://www.bouldercounty.org/env/sustainability/pages/sustainableenergyplan.aspx. Accessed 10/18/2013. .

BUSD. 2007. *Resolution Number 07-52.* Available: http://www.berkeley.k12.ca.us/SB/docs/bd_of_ed_june_20_07_packet.pdf. Accessed 9/28/2007.

Butler. 2008. *City's Planning Board Finalizes its Wind Energy Ordinance. Sentinel and Enterprise. 12 January 2008.* Available: http://www.sentinelandenterprise.com/mobile/ci_7953816. Accessed 1/28/2008.

California Assembly. 2001. *Chapter 562: An Act Relating to Wind Energy.* Available: http://www.leginfo.ca.gov/pub/01-02/bill/asm/ab_1201-1250/ab_1207_bill_20011007_chaptered.pdf. Accessed 7/6/2007.

California Public Utilities Commission. 2013. *About the California Solar Initiative.* Available: http://www.cpuc.ca.gov/puc/energy/solar/aboutsolar.htm. Accessed 8/30/2013.

Carolina Live. 2012. *More energy created by wind in North Myrtle Beach.* Available: http://www.carolinalive.com/news/story.aspx?id=722048#.UmA_qHCkqN0. Accessed 10/17/2013.

Cayuga County Soil and Water Conservation District. 2013. *Anaerobic Digestion at the Cayuga County Soil and Water Conservation District's Community Digester.* Available: http://www.cayugaswcd.org/digester.html. Accessed 10/17/2013.

CEC. 2007. *DER Equipment: Microturbines.* Available: http://www.energy.ca.gov/distgen/equipment/microturbines/cost.html. Accessed 10/1/2007.

CEC. 2007a. *DER Equipment: Reciprocating Engines.* Available: http://www.energy.ca.gov/distgen/equipment/reciprocating_engines/cost.html. Accessed 10/1/2007.

Chevron. 2007. *From Waste to Watts: City of Rialto Teams with Chevron, Fuel Cell Energy to Turn Restaurant Grease into Renewable Power.* Available: http://www.chevron.com/news/press/2007/2007-05-08.asp. Accessed 9/28/2007.

City of Auburn and Cayuga County. 2009. *Comprehensive Sustainable Energy and Development Plan.* Available: http://www.auburnny.gov/public_documents/auburnny_planning/Energy_Master_Plan.pdf. Accessed 9/4/2013.

City of Boulder. 2013. *Energy Efficiency Upgrades at City Facilities – Energy Performance Contract.* Available: https://bouldercolorado.gov/public-works/energy-efficiency-upgrades-at-city-facilities-energy-performance-contract-epc. Accessed 8/30/2013.

City of Milwaukee. 2013. *Milwaukee Shines.* Available: http://city.milwaukee.gov/milwaukeeshines. Accessed 8/29/2013.

Clean Energy Jobs Act. 2012. *Voters Approve Prop 39 and Implementing Legislation Passes.* Available: http://www.cleanenergyjobsact.com/. Accessed 8/28/2013.

Columbia Water & Light. 2013. *2013 Renewable Energy Report.* Available: http://www.gocolumbiamo.com/WaterandLight/Documents/RenewReport.pdf. Accessed 8/29/2013.

The Daily Caller. 2012. *Successes, Failures For Major Green Energy Ballot Initiatives.* Available: http://dailycaller.com/2012/11/07/successes-failures-for-major-green-energy-ballot-initiatives/2/. Accessed 8/29/2013.

DSIRE. 2007. *Lakeland Electric – Solar Water Heating Program.* Available: http://www.dsireusa.org/library/includes/incentive2.cfm?Incentive_Code = FL51F&state = FL&CurrentPageID = 1&RE = 1&EE = 1. Accessed 7/9/2007.

DSIRE. 2007a. *Pike County – Wind Turbine Siting Standards.* Available: http://www.dsireusa.org/library/includes/incentive2.cfm?Incentive_Code = IL08R&state = IL&CurrentPageID = 1&RE = 1&EE = 0. Accessed 7/9/2007.

DSIRE. 2007b. *Rockingham County – Small Wind Ordinance.* Available: http://dsireusa.org/library/includes/incentivesearch.cfm?Incentive_Code = VA07R&Search = TableType&type = Constr&CurrentPageID = 7&EE = 1&RE = 1. Accessed 7/18/2007.

DSIRE. 2007c. *Yellow Springs Utilities – Net Metering.* Available: http://www.yso.com/ordinances/pdf/1042.pdf. Accessed 7/18/2007.

DSIRE. 2007d. *City of New Orleans – Net Metering.* Available: http://www.dsireusa.org/library/includes/incentive2.cfm?Incentive_Code = LA05R&state = LA&CurrentPageID = 1&RE = 1&EE = 1. Accessed 7/19/2007.

DSIRE. 2012. *City of Tallahassee Utilities – Solar Water Heating Rebate.* Available: http://dsireusa.org/incentives/incentive.cfm?Incentive_Code=FL57F&re=0&ee=0. Accessed 8/29/2013.

DSIRE. 2012a. *Maryland – Commercial Clean Energy Grant Program.* Available: http://www.dsireusa.org/incentives/incentive.cfm?Incentive_Code=MD47F&re=1&ee=1. Accessed 8/30/2013.

DSIRE. 2013. *Net Metering.* Available: http://www.dsireusa.org/solar/solarpolicyguide/?id=17. Accessed 8/29/2013.

DSIRE. 2013a. *California Financial Incentives.* Available: http://www.dsireusa.org/incentives/index.cfm?re=0&ee=0&spv=0&st=0&srp=1&state=CA. Accessed 8/29/2013.

DSIRE. 2013b. *Columbia – Renewables Portfolio Standard.* Available: http://www.dsireusa.org/incentives/incentive.cfm?Incentive_Code=MO04R&ee=0. Accessed 8/29/2013.

DSIRE. 2013c. *Harford County – Property Tax Credit for Solar and Geothermal Devices.* Available: http://www.dsireusa.org/solar/incentives/incentive.cfm?Incentive_Code=MD24F&re=0&ee=0. Accessed 8/30/2013.

DSIRE. 2013d. *Oregon – Small-Scale Energy Loan Program.* Available: http://www.dsireusa.org/incentives/incentive.cfm?Incentive_Code=OR04F. Accessed 8/30/2013.

DSIRE. 2013e. *Alaska – Renewable Energy Grant Program.* Available: http://www.dsire-usa.org/incentives/incentive.cfm?Incentive_Code=AK12F&re=1&ee=1. Accessed 8/30/2013.

DSIRE. 2013f. *Massachusetts – Model As-of Right Zoning Ordinance or Bylaw: Allowing Use of Wind Energy Facilities.* Available: http://www.dsireusa.org/incentives/incentive.cfm?Incentive_Code=MA14R&re=0&ee=0. Accessed 8/30/2013.

EIA. 2003. *Biomass for Electricity Generation.* Available: http://www.eia.doe.gov/oiaf/analysispaper/biomass/table7.html. Accessed 10/1/2007.

EIA. 2013. *Electricity Monthly Update, August 2013.* Available: http://www.eia.gov/electricity/monthly/update/end_use.cfm. Accessed 8/29/2013.

EIA. 2013a. *Electricity Monthly Update, August 2013.* Available: http://www.eia.gov/electricity/monthly/update/end_use.cfm#tabs_prices-3. Accessed 8/29/2013.

Ellensburg. 2007. *Ellensburg Solar Community.* Available: http://www.solarwashington.org/Calendar/events/2007/ESC-Overview.pdf. Accessed 7/18/2007.

Energy Services Bulletin. 2004. *Lenox wind turbine generates power, interest in renewable energy.* Available: http://www.wapa.gov/ES/pubs/esb/2004/feb/feb043.htm. Accessed 1/29/2008.

Gang, D. W. 2007. *San Bernardino County Vows to Protect Environment. Press-Enterprise, 28 August, 2007.* Available: http://www.pe.com/. Accessed 1/29/2008.

GeoExchange. 1997. *North Bonneville City Hall, North Bonneville, Washington.* Available: http://www.geoexchange.org/pdf/cs-056.pdf. Accessed 7/17/2007.

Green Jobs. 2007. *Methane Generator, Energy Conservation Benefit the City of Sparks.* Available: http://www.greenjobs.com/Public/IndustryNews/inews02542.htm. Accessed 7/6/2007.

Hayward. 2005. *Resolution Authorizing the Installation of a Solar Power Electrical Generating System Atop the City's Barnes Court Warehouse.* Available: http://www.ci.hayward.ca.us/citygov/meetings/cca/rp/2005/rp022205-03.pdf. Accessed 7/10/2007.

Highland Beach. 2006. *Highland Beach "Green" Town Hall.* Available: http://www.dnr.state.md.us/ed/pdfs/HighlandBeach.pdf. Accessed 3/12/2008.

Honeywell. 2006. *Solar Canopy Generates Savings, Lightens Tax Burden.* Available: https://buildingsolutions.honeywell.com/NR/rdonlyres/E008D8D8-B91E-4153-BD85-5AC8BC4CDC10/59152/e008d8d8b91e4153bd855ac8bc4cdc10.pdf. Accessed 1/30/2008.

Hull. 2008. *Hull Wind.* Available: http://www.hullwind.org/. Accessed 1/29/2008.

Hull Municipal Light Plant. 2013. *An Analysis of Wind Power Development in the Town of Hull, MA.* Available: https://www.google.com/url?sa=t&rct=j&q=&esrc=s&source=web&cd=4&ved=0CEQQFjAD&url=http%3A%2F%2Fwww.town.hull.ma.us%2FPublic_Documents%2FHullMA_Light%2Ffinaloffshorewind%2FFinal%2520Report.docx&ei=ZFgSUr6LBIP42QWf5oGwCQ&usg=AFQjCNGbwTEoXquEvTx535ghXJsS6ScvjQ&sig2=vPjLLwpQSOjkgB3eFr8CnQ. Accessed 8/29/2013.

Hunterdon County Democrat. 2012. *Bayonne School Stays in Power After Sandy: Thanks to a Solar System from Flemington Company.* Available: http://nj.com/hunterdon-county-democrat/index.ssf/2012/11/bayonne_school_stays_in_power.html. Accessed 9/11/2013.

Hydropower Reform Coalition. 2009. *Skagit River Project.* Available: http://www.hydroreform.org/sites/default/files/Skagit_FINAL_1_0.pdf. Accessed 8/29/2013.

ICLEI. Undated. *Case Study: Waverly, Iowa.* Available: http://www.greenpowergovs.org/wind/Waverly%20case%20study.html. Accessed 7/19/2007.

ICLEI. 2013. *River Falls, Wisconsin.* Available: http://www.icleiusa.org/action-center/river-falls-wi-case-study-ncsc. Accessed 8/29/2013.

ICECF. 2013. *About the Foundation.* Available: http://www.illinoiscleanenergy.org/about-the-foundation/. Accessed 8/30/2013.

IREC. 2007. *Net Metering.* Available: http://www.dsireusa.org/documents/SummaryMaps/NetMetering_Map.ppt. Accessed 7/25/2007.

IREC. 2008. *Solar Boston: Creating a Citywide Solar Strategy for the City of Boston.* Available: http://www.irecusa.org/index.php?id = 68&tx_ttnews[pS] = 1203364584&tx_ttnews[tt_news] = 914&tx_ttnews[backPid] = 70&cHash = 8da3f996b1. Accessed 3/11/2008.

Lazard. 2012. *Levelized Cost of Energy Analysis – Version 6.0.* Available: https://www.misoenergy.org/Library/Repository/Meeting%20Material/Stakeholder/PAC/2012/20121221/20121221%20PAC%20Supplemental%20Levelized%20Cost%20of%20Energy%20Analysis.pdf.

LBNL. 2002. *Berkeley Lab and the Clean Energy Group: Case Studies of State Support for Renewable Energy.* Available: http://eetd.lbl.gov/ea/ems/cases/Bulk_Purchases.pdf. Accessed 1/29/2008.

LIHI. 2008. *Low-Impact Hydropower Institute.* Available: http://lowimpacthydro.org/cf.aspx. Accessed 1/29/2008.

LIHI. 2008a. *LIHI Certificate #5 – Skagit Project, Skagit River, Washington, (FERC #553).* Available: http://www.lowimpacthydro.org/lihi-certificate-5-skagit-project-skagit-river-washington-ferc-553.html. Accessed 8/29/2013.

Maricopa Association of Governments. 2002. *Permit Submittal Requirements for PV Systems.* Available: http://www.mag.maricopa.gov/pdf/cms.agendas/BCMay15attach_361.pdf. Accessed 7/19/2007.

Marin. 2008. *Marin County Code, Section 20.20.030 Energy Conservation Code.* Available: http://municipalcodes.lexisnexis.com/codes/marincounty/_DATA/TITLE20I/Chapter_20_20_DESIGN_AND_IMPRO.html#2. Accessed 1/28/2008.

Mason City. 2006. *Zoning for Wind Energy Conversion Systems.* Available: http://www.windustry.org/SmallWind/documents/MasonCityIAWindOrdinance.pdf. Accessed 7/19/2007.

McQuay International. 2003. *Case Study: Auburn Preserves City Hall, Improves Employee Comfort With Innovative Geoexchange System.* Available: http://www.daikinmcquay.com/mcquaybiz/marketing_tools/mt_at_wshp/RefJob/AuburnFinal.pdf. Accessed 8/29/2013.

MDER. 2007. *Model Amendment to a Zoning Ordinance or By-Law: Allowing Wind Facilities by Special Permit.* Available: http://www.mass.gov/Eoca/docs/doer/renew/model-allow-wind-by-permit.pdf. Accessed 1/29/2008.

Michigan DLEG. 2007. *Michigan Siting Guidelines for Wind Energy Systems.* Available: http://www.michigan.gov/documents/Wind_and_Solar_Siting_Guidlines_Draft_5_96872_7.pdf. Accessed 7/6/2007.

MTC. 2002. *Power Quality Problems and Renewable Energy Solutions.* Available: http://www.mtpc.org/Project%20Deliverables/PP_General_Power_Quality_Study.pdf. Accessed 7/9/2007.

New Jersey. 2006. *NJBPU Unveils Largest East Cost Solar Project.* Available: http://www.njcleanenergy.com/html/5library/press/pr_njbpu-largesolar.html. Accessed 7/6/2007.

Newton. 2005. *Energy Action Plan.* Available: http://www.greenenergynewton.org/sunergy/EAP021005.pdf. Accessed 9/28/2007.

NREL. 2007. *Solar Process Heating.* Available: http://www.nrel.gov/learning/re_solar_process.html. Accessed 7/6/2007.

NREL. 2007a. *Solar Hot Water.* Available: http://www.nrel.gov/learning/re_solar_hot_water.html. Accessed 7/6/2007.

NREL. 2007b. *Innovations in Renewable Energy Financing.* Available: http://www.usaee.org/usaee2007/submissions/OnlineProceedings/Innovations%20in%20Renewable%20Energy%20Financing%20_Cory_%20-%20FINAL.pdf. Accessed 1/30/2008.

NREL. 2011a. *Solar Schools Assessment and Implementation Project: Financing Options for Solar Installations on K–12 Schools.* Available: http://www.nrel.gov/docs/fy12osti/51815.pdf. Accessed 8/30/2013.

NREL. 2011b. *Economic Impacts from the Boulder County, Colorado, ClimateSmart Loan Program: Using Property-Assessed Clean Energy (PACE) Financing.* Available: http://www.nrel.gov/docs/fy11osti/52231. pdf. Accessed 9/11/2013.

NREL. 2012. *Preliminary Analysis of the Jobs and Economic Impacts of Renewable Energy Projects Supported by the $1603 Treasury Grant Program.* Available: http://www.nrel.gov/docs/fy12osti/52739.pdf. Accessed 8/29/2013.

NREL. 2012a. *Renewable Electricity Futures Study.* Available: http://en.openei.org/apps/TCDB/. Accessed 8/30/2013.

NYSERDA. 2007. *On-site or Small Wind Incentives.* Available: http://www.powernaturally.org/Programs/ Wind/incentives.asp?i = 8. Accessed 7/5/2007.

NYSERDA. 2013. *PON 2112 - Solar PV Program Financial Incentives.* Available: http://www.nyserda. ny.gov/Funding-Opportunities/Current-Funding-Opportunities/PON-2112-Solar-PV-Program-Financial-Incentives.aspx?sc_database=web. Accessed 8/30/2013.

Pennsylvania. 2006. *Governor Rendell Announces Scholarships for Local Government Officials to Attend Wind Energy Conference.* Available: http://www.state.pa.us/ papower/cwp/view.asp?Q = 452210&A = 11. Accessed 7/6/2007.

Phoenix. 2007. *Solar and Other Renewable Energy Projects.* Available: http://www.phoenix.gov/sustainability/solarproj.html. Accessed 7/19/2007.

Portland. 2007. *Powered by the Sun and Wind!* Available: http://www.portlandonline.com/osd/index. cfm?c = 42399&a = 117668. Accessed 7/19/2007.

Portsmouth. 2007. *Wind Project for the High School and Middle School Feasibility Study.* Available: http:// www.portsmouthrienergy.com/schoolstudy9-07.htm. Accessed 9/27/2007.

PR Newswire. 2011. *Nexterra Receives $6.9 Million Order to Deliver Biomass Gasification System to Department of Veterans Affairs Medical Center in Michigan.* Available: http://www.prnewswire.com/news-releases/ nexterra-receives-69-million-order-to-deliver-biomass-gasification-system-to-department-of-veterans-affairs-medical-center-in-michigan-129710358. html. Accessed 8/29/2013.

Renewable Energy World. 2013. *What Will Replace the California Solar Initiative?* Available: http:// www.renewableenergyworld.com/rea/news/article/2013/07/what-will-replace-the-california-solar-initiative. Accessed 8/30/2013.

REPP-CREST. Undated. *Geothermal Resources.* Available: http://www.crest.org/geothermal/geothermal_ brief_economics.html. Accessed 10/1/2007.

Ryan, A. 2007. *Federal Money Charges Portland Solar Effort.* Available: http://www.djc-or.com/viewStory. cfm?recid = 29664&userID = 1. Accessed 7/17/2007.

San Diego. 2007. *City of San Diego Unveils 1-Megawatt Solar System at City's Alvarado Water Treatment Plant.* Available: http://www.renewableenergyaccess.com/ rea/partner/story?id = 47619. Accessed 1/31/2008.

Sandia. 2007. *Technical Issues Concerning Third Party Financing for Renewable Energy.* Sandia National Laboratories. Available: http://energy.sandia.gov/ technicalissues.htm. Accessed 1/30/2008.

San Francisco. 2007. *Mayor Newsome Announces Plan to Significantly Expand Solar, Renewable Energy Generation in San Francisco.* Available: http://sfgov.org/site/ mayor_index.asp?id = 55941. Accessed 10/1/2007.

San Francisco. 2007a. *Mayor Newsom Unveils SFO/ SFPUC Solar Energy Project.* Available: http://sfgov. org/site/mayor_index.asp?id = 68140. Accessed 10/1/2007.

San Jose. 2009. *Memorandum of Understanding.* Available: http://www3.sanjoseca.gov/clerk/ Agenda/20090616/20090616_0701.pdf. Accessed 8/30/2013.

Santa Monica. 2007. *Solar and Renewable Energy.* Available: http://www.smgov.net/epd/residents/ Energy/solar.htm. Accessed 7/20/2007.

SDREO. 2007. *How much electricity will PV produce?* Available: http://www.sdreo.org/ContentPage. asp?ContentID = 121&SectionID = 86&SectionTarget = 44. Accessed 9/28/2007.

Seattle Green Power. 2007. *Meridian Park Elementary School.* Available: http://www.ci.seattle.wa.us/light/ Green/greenPower/Accomplishments/meridian.asp. Accessed 9/28/2007.

Sioux City Journal. 2012. *Spirit Lake, Iowa, School District Honored For Wind Turbines.* Available: http://siouxcityjournal.com/news/local/spirit-lake-iowa-school-district-honored-for-wind-turbines/ article_055c6490-cb1f-51f8-b47e-6681825de968. html. Accessed 8/29/2013.

SMUD. 2007. *Folsom-SMUD partnership makes solar easier, more affordable.* Available: http://www.smud. org/news/releases/07archive/11_28_07_smudfolsom-solar.pdf. Accessed 3/11/2008.

Stimmel, R. 2007. *AWEA Small Wind Turbine Global Market Study. American Wind Energy Association.* Available: http://www.awea.org/smallwind/docu-ments/AWEASmallWindMarketStudy2007.pdf. Accessed 7/25/2007.

Tallahassee. 2007. *Solar Power Net Metering.* Available: http://www.talgov.com/you/electric/net_metering. cfm. Accessed 1/30/2008.

Tampa. 2007. *Florida's Renewable Energy Source.* Available: http://www.energyvortex.com/pages/ headlinedetails.cfm?id = 1304&archive = 1. Accessed 1/29/2008.

Tucson. 2005. *Resolution 20193.* Available: http://www. dsireusa.org/documents/Incentives/AZ26F.htm. Accessed 7/25/2007.

U.S. Conference of Mayors. 2007. *Climate Protection Strategies and Best Practices Guide.* Available: http://www.usmayors.org/climateprotection/ documents/2007bestpractices-mcps.pdf. Accessed 4/8/2008.

U.S. DOE. 2003. *Wind Power Pioneer Interview: Glenn Cannon, Waverly Light and Power.* Available: http:// www.eere.energy.gov/windandhydro/windpower-ingamerica/filter_detail.asp?itemid = 688. Accessed 7/19/2007.

U.S. DOE. 2005. *Small Wind Electric Systems: A U.S. Consumer's Guide.* Available: http://www.eere. energy.gov/windandhydro/windpoweringamerica/ pdfs/small_wind/small_wind_guide.pdf. Accessed 7/5/2007.

U.S. DOE. 2006. *Geothermal Heat Pumps.* Available: http://www1.eere.energy.gov/geothermal/heatpumps. html. Accessed 7/9/2007.

U.S. DOE. 2006a. *Net Metering.* Available: http://www. eere.energy.gov/greenpower/markets/netmetering. shtml. Accessed 7/25/2007.

U.S. DOE. 2007. *Annual Report on U.S. Wind Power Installation, Cost, and Performance Trends: 2006.* Available: http://www1.eere.energy.gov/windandhydro/ pdfs/41435.pdf. Accessed 9/27/2007.

U.S. DOE. 2007a. *California Cities and Counties Reduce or Eliminate Solar Permit Fees.* Available: http://www. eere.energy.gov/state_energy_program/project_brief_ detail.cfm/pb_id = 1179. Accessed 1/29/2008.

U.S. DOE. 2008. *Department of Energy Selects Winner of Wind Cooperative of the Year Award.* Available: http:// www.energy.gov/news/6010.htm. Accessed 3/11/2008.

U.S. DOE. 2010. *2009 Renewable Energy Data Book.* Available: http://www1.eere.energy.gov/maps_data/ pdfs/eere_databook.pdf. Accessed 9/4/2013.

U.S. DOE. 2011. *Solar in Action.* Available: http:// www1.eere.energy.gov/solar/pdfs/50192_annarbor. pdf. Accessed 9/4/2013.

U.S. DOE. 2012. *Spirit Lake School District Case Study.* Available: http://www.windpoweringamerica.gov/ filter_detail.asp?itemid=3623. Accessed 8/29/2013.

U.S. DOE. 2012a. *Columbia – Renewables Portfolio Standard.* Available: http://energy.gov/savings/columbia-renewables-portfolio-standard. Accessed 8/29/2013.

U.S. DOE. 2013. *Solar For Milwaukee, By Milwaukee.* Available: http://energy.gov/articles/solar-milwaukee-milwaukee. Accessed 8/29/2013.

U.S. DOE. 2013a. *2012 Wind Technologies Market Report.* Available: http://www1.eere.energy.gov/wind/pdfs/2012_wind_technologies_market_report.pdf. Accessed 8/30/2013.

U.S. EPA. 2000. *Biomass Energy.* Available: http://yosemite.epa.gov/oar/globalwarming.nsf/UniqueKey-Lookup/SHSU5BNJXH/$File/biomassenergy.pdf. Accessed 7/6/2007.

U.S. EPA. 2003. *Financing Energy Efficiency Projects. Government Finance Review, February 2003.* Available: http://www.energystar.gov/ia/business/government/Financial_Energy_Efficiency_Projects.pdf. Accessed 5/29/2007.

U.S. EPA. 2004. *Guide to Purchasing Green Power.* Available: http://www.epa.gov/greenpower/buygreen-power/guide.htm. Accessed 7/6/2007.

U.S. EPA. 2004a. *A Comparison of Dairy Cattle Manure Management with and without Anaerobic Digestion and Biogas Utilization.* Available: http://www.epa.gov/agstar/documents/nydairy2003.pdf. Accessed 9/26/2007.

U.S. EPA. 2006a. *Electricity from Hydropower.* Available: http://www.epa.gov/cleanrgy/hydro.htm. Accessed 7/9/2007.

U.S. EPA. 2006b. *Clean Energy-Environment Guide to Action.* Available: http://www.epa.gov/statelocalclimate/documents/pdf/guide_action_full.pdf. Accessed 7/19/2007.

U.S. EPA. 2007. *Green Power Partnership: Green Power Defined.* Available: http://www.epa.gov/greenpower/gpmarket/index.htm. Accessed 4/8/2008.

U.S. EPA. 2007a. *Green Power Partnership: Glossary.* Available: http://www.epa.gov/greenpower/pubs/glossary.htm. Accessed 4/8/2008.

U.S. EPA. 2008. *Air Emissions.* Available: http://www.epa.gov/cleanenergy/energy-and-you/affect/air-emissions.html. Accessed 3/28/2008.

U.S. EPA. 2011. *Overview of Greenhouse Gases.* Available: http://epa.gov/climatechange/ghgemissions/gases/ch4.html. Accessed 8/29/2013.

U.S. EPA. 2012. *Solar Energy.* Available: http://www.epa.gov/region1/eco/energy/re_solar.html. Accessed 10/17/2013.

U.S. EPA. Undated. *Green Power from Landfill Gas.* Available: http://www.epa.gov/lmop/docs/LMOP-GreenPower.pdf. Accessed 9/28/2007.

Waste Management World. 2013. *90,000 TPA Dry Anaerobic Digestion Biogas Plant in San Jose Uses Special Concrete.* Available: http://www.waste-management-world.com/articles/2013/08/90-000-tpa-dry-anaerobic-digestion-biogas-plant-in-san-jose-uses-special-concrete.html. Accessed 8/30/2013.

Waverly. 2007. *City of Waverly.* Available: http://city.waverlyia.com/. Accessed 7/19/2007.

WBDG. 2007. *Fuel Cells.* Available: http://www.wbdg.org/resources/fuelcell.php. Accessed 10/1/2007.

WLP. 2005. *Greenhouse Gas Report.* Available: http://wlp.waverlyia.com/docs/Copy%20of%20WLP-2005Report.pdf. Accessed 7/19/2007.

WLP. 2006. *Renewables Production 2006.* Available: http://wlp.waverlyia.com/docs/Renewables%20file.pdf. Accessed 7/19/2007.

WLP. 2007. *Waverly Light and Power Purchases Two Wind Turbines.* Available: http://wlp.waverlyia.com/headline.asp?obj_id = 1092. Accessed 7/19/2007.

WLP. 2007a. *Renewable Energy.* Available: http://wlp.waverlyia.com/renewable_energy.asp. Accessed 7/19/2007.

WRI. 2007. *The Solar Services Model: An Innovative Financing Approach to On-site Solar Photovoltaics.* Available: http://www.thegreenpowergroup.org/pdf/case_studies_Staples_2.pdf. Accessed 1/31/2008.

Zero Waste Energy. Undated. *San Jose Anaerobic Digestion/Composting Plant.* Available: http://www.zerowasteenergy.com/content/san-jose-anaerobic-digestioncomposting-plant. Accessed 8/30/2013.

www.ingramcontent.com/pod-product-compliance
Lightning Source LLC
Chambersburg PA
CBHW081407170526
45166CB00010B/3248

* 9 7 8 1 5 0 0 3 0 9 1 0 7 *